DESIGNING WITH FIELD-EFFECT TRANSISTORS

DESIGNING WITH FIELD-EFFECT TRANSISTORS

SILICONIX INC.

Editor in Chief

Arthur D. Evans

McGRAW-HILL BOOK COMPANY

*New York   St. Louis   San Francisco   Auckland   Bogotá   Hamburg
Johannesburg   London   Madrid   Mexico   Montreal   New Delhi
Panama   Paris   São Paulo   Singapore   Sydney   Tokyo   Toronto*

3 3001 00597 5478

Library of Congress Cataloging in Publication Data

Siliconix Incorporated.
Designing with field-effect transistors.

Includes index.
1. Field-effect transistors. 2. Transistor cir-
cuits. 3. Electronic circuit design. I. Evans,
Arthur D. II. Title.
TK7871.95.S54 1981   621.3815'284   80–15690
ISBN 0–07–057449–9

2 3 4 5 6 7 8 9 0 KPKP 8 9 8 7 6 5 4 3 2 1

The editors for this book were Tyler G. Hicks and Geraldine
Fahey, the designer was Elliot Epstein, and the production
supervisor was Sally Fliess. It was set in Baskerville by
The Kingsport Press.

Printed and bound by The Kingsport Press.

CONTENTS

Preface
ix

1
FIELD-EFFECT TRANSISTOR THEORY
1

2
PARAMETERS AND SPECIFICATIONS
25

3
LOW-FREQUENCY CIRCUITS
61

4
HIGH-FREQUENCY CIRCUITS
137

5
ANALOG SWITCHES
193

6
**VOLTAGE-CONTROLLED RESISTORS
AND FET CURRENT SOURCES**
233

7
POWER FETs
255

8
FETs IN INTEGRATED CIRCUITS
281

INDEX
289

v

CONTRIBUTORS

Written by the Applications Engineering Staff
of Siliconix Incorporated

Editor in Chief

ARTHUR D. EVANS

Contributors

Arthur D. Evans	(Chapters 1, 2, and 5)
Walt Heinzer	(Chapters 3 and 6)
Ed Oxner	(Chapter 4)
Lee Shaeffer	(Chapters 7 and 8)

PREFACE

The purpose of this book is to aid the electronics circuit designer in the utilization of the field-effect transistor (FET). Since its emergence from the development laboratory in about 1962, the FET has become an important and widely used component in the electronic industry.

With its importance increasing annually, both as a discrete component and in integrated circuits, it is essential that the serious circuit designer have an insight into how the FET behaves under different circuit and environmental conditions. It is also helpful to understand how its physical and electrical characteristics are interrelated. This book goes just deep enough into FET theory to provide that insight.

After the introductory chapter describing how the device works, the interrelationship of the various FET characteristics and how they relate to typical data sheet specifications is discussed.

Subsequent chapters deal with various categories of circuit applications of the FET. The order of these application chapters is rather arbitrary; each chapter stands on its own. Detailed circuit design examples are given to illustrate applications.

As this book was being readied for publication, the availability of FETs designed for power applications was expanding rapidly. The chapter on Power FETs spotlights this relatively new field.

Art Evans

1

FIELD-EFFECT
TRANSISTOR THEORY

1-1 Introduction
1-2 The Junction FET
1-3 The MOSFET (Insulated-Gate FET)
1-4 FET Symbols
1-5 Physical Characteristics of FETs
1-6 Summary

1-1 INTRODUCTION

This handbook is designed principally for the FET user. The primary emphasis has been placed on allowing the reader to learn necessary information quickly, without becoming bogged down in complicated analyses. The first two chapters provide sufficient background information to answer many questions asked by FET circuit design engineers. The aim of these chapters is to give design engineers an intuitive sense of how to manipulate the circuits presented in the remaining chapters, which contain both theory and "cookbook" applications. Some knowledge by the reader of semiconductor theory is assumed. A prior understanding of the concept of conduction by electrons and holes, of the "forward" and "reverse" characteristics of a p-n junction, and of the "depletion" region at a p-n junction will be helpful in understanding FET theory.

There are two general types of transistors: bipolar and unipolar. The unipolar, more commonly called the "field-effect transistor" (FET), is the subject of this book. The concept of controlling the electronic conduction in a solid by an electric field predates the invention of the bipolar transistor. J. E. Lilienfeld filed for a patent on such a device in 1925, as shown in Fig. 1-1. W. Shockley presented a comprehensive theory

Patented Jan. 28, 1930

1,745,175

UNITED STATES PATENT OFFICE

JULIUS EDGAR LILIENFELD, OF BROOKLYN, NEW YORK

METHOD AND APPARATUS FOR CONTROLLING ELECTRIC CURRENTS

Application filed October 8, 1926. Serial No. 140,363, and in Canada October 22, 1925.

The invention relates to a method of and apparatus for controlling the flow of an electric current between two terminals of an electrically conducting solid by establishing a
5 third potential between said terminals; and is particularly adaptable to the amplification of oscillating currents such as prevail, for example, in radio communication. Heretofore, thermionic tubes or valves have been
10 generally employed for this p⎯
the present invention has fo⎯
pense entirely with dev⎯
transmission of el⎯
space and e⎯
15 acter wh⎯
an i⎯
ha⎯

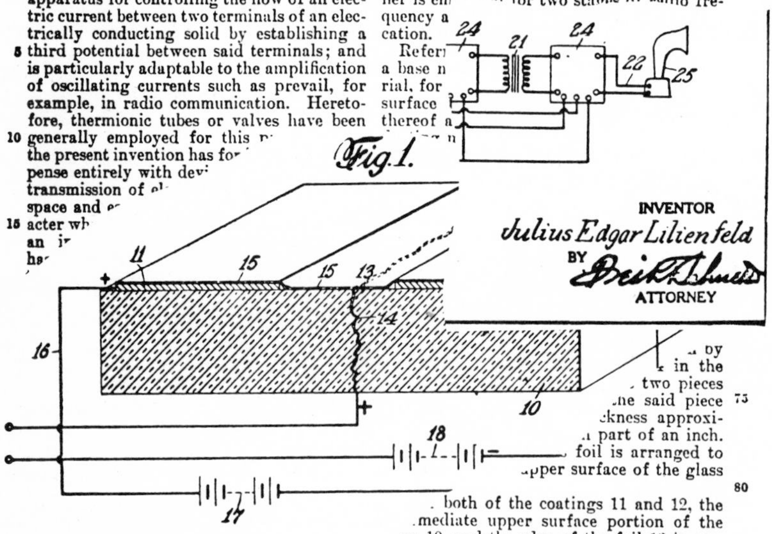

receiving circuit in which the novel amplifier is em⎯ ⎯ for two stages of radio frequency a⎯
cation.

Refer⎯
a base n⎯
rial, for⎯
surface f⎯
thereof a⎯

Fig. 1.

INVENTOR
Julius Edgar Lilienfeld
BY
Beik L. Shueli
ATTORNEY

⎯ by ⎯ in the ⎯ two pieces
75 ⎯ne said piece ⎯ckness approxi⎯ ⎯ part of an inch.
⎯ foil is arranged to ⎯pper surface of the glass
80 ⎯ both of the coatings 11 and 12, the ⎯mediate upper surface portion of the ⎯ass 10, and the edge of the foil 13 is provided a film or coating 15 of a compound
85 having the property of acting in conjunction with said metal foil electrode as an element of uni-directional conductivity. That is to say, this coating is to be electrically conductive and possess also the property, when
90 associated with other suitable conductors, of establishing at the surface of contact a considerable drop of potential. The thickness of the film, moreover, is minute and of such a degree that the electrical conductivity
95 therethru would be influenced by applying thereto an electrostatic force. A suitable material for this film and especially suitable in conjunction with aluminum foil, is a compound of copper and sulphur. A convenient
100 way of providing the film over the coatings

35 ing means at a predetermined potential
which is to be substantially in excess of a
pote⎯
circ⎯   *Although the word "TRANSISTOR" had*
Th⎯   rill
40 best   *not been coined yet, it appears that the*   ⎯ec-
tion   *"Field Effect" Transistor concept pre-*   in
whic⎯
Fi⎯   *dates the invention of the Bipolar*   ⎯tly
enla⎯   the
45 nove   *Junction Transistor by a few years.*   of
exan⎯
Fi⎯   *(say about 23 years!)*   ing
the voltage characteristics of an amplifier
as shown in Fig. 1.
50   Fig. 3 is a diagrammatic view of a radio

FIGURE 1-1   "FET" patent.

of the field-effect transistor in 1952.[2] However, the commercial availability of the FET in the early 1960s followed the bipolar transistor by 8 or 10 years. The superior performance of the FET in many circuit applications previously utilizing bipolar transistors or vacuum tubes resulted in a rapid growth in its acceptance as an important electronic component.

Several types of semiconductor materials have been used for making FETs: silicon, germanium, gallium arsenide, and others. By far the most widely used is silicon, and unless otherwise specified all device types discussed in this book are silicon types.

The field-effect transistor (FET) is a class of electronic semiconductor device in which the conduction of a "channel" between source (S) and drain (D) terminals is controlled by an electric field impressed upon the channel via a gate (G) terminal. The conducting channel may utilize n-type carriers (electrons) or p-type carriers (holes). The electric field which controls the channel conduction may be introduced via a p-n junction (for a "junction" FET), a metal plate separated from the semiconductor channel by an oxide dielectric (for a metal-oxide-semiconductor FET), or a combination of the two methods. The polarity of the controlling electric field is a function of the type of carriers in the channel. A FET "family tree" is shown in Fig. 1-2.

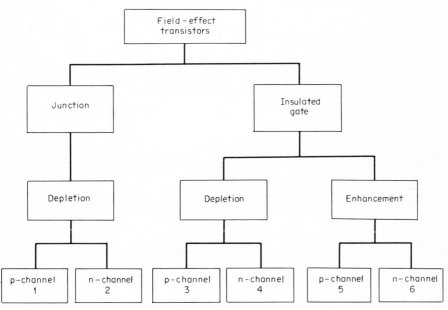

FIGURE 1-2 FET family tree.

1-2   THE JUNCTION FET

A section view of a junction field-effect transistor (JFET) is shown in Fig. 1-3. This structure contains, between the source and the drain contacts, an n-type "channel" embedded in a p-type silicon substrate. If it is assumed that the p-n junction forms a barrier to current flow, then it can be seen that channel conduction is a function of the channel width, length, and thickness and of the density and mobility of the carriers. In this structure current can flow equally well in either direction through the channel; i.e., the drain may be positive or negative with respect to the source.

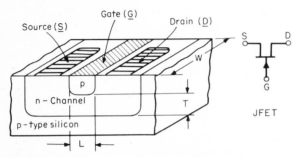

FIGURE 1-3   Junction-type field-effect transistor.

1-2-1   Channel Conductance (g_{ds})

We will examine how the channel conduction is controlled by a gate-to-channel voltage. As indicated in Fig. 1-4a, at the p-n (gate-to-channel) junction there is a depletion layer. If there is an abrupt change from a high concentration of holes in the gate region (p-type) to a much lower concentration of electrons in the channel region (n-type), then most of the depletion width will occur in the channel. The depletion width is proportional to the square root of the junction potential.[2]

$$W_d = \left(\frac{V_{bi} \pm V_{GS}}{K_1 N_c}\right)^{1/2}$$

where W_d = junction depletion width
$\quad\quad V_{bi}$ = built-in junction potential
$\quad\quad V_{GS}$ = reverse or forward bias applied to gate-channel junction
$\quad\quad K_1$ = constant
$\quad\quad N_c$ = channel carrier concentration

The channel conduction is decreased as the junction depletion width is increased and thus can be controlled by the gate-source voltage V_{GS}.

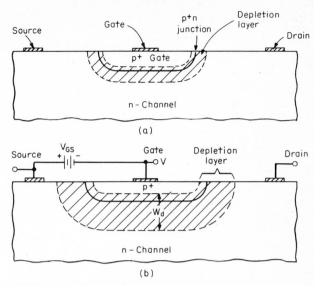

FIGURE 1-4   Depletion layer at p^+n junction.

The channel conduction g_{ds} of a typical JFET as a function of the junction voltage is shown in Fig. 1-5. The characteristic is shown with zero volts between the "source" and the "drain"; thus the depletion thickness is uniform along the channel length. The voltage at which the channel conduction approaches zero is the "gate cutoff voltage," and is given the symbol $V_{GS(off)}$. The polarity of $V_{GS(off)}$ is such that the p-n gate-channel junction is *reverse*-biased. A forward bias at the gate junction will increase channel conduction because of the resulting decrease in depletion layers; however, as indicated in Fig. 1-6, a *forward* gate-channel bias results in a large increase in gate current. In the FET

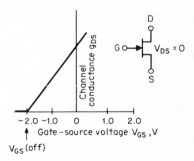

FIGURE 1-5   Channel conductance vs. gate-channel voltage.

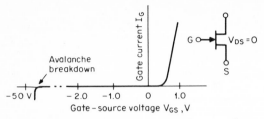

FIGURE 1-6   Gate current vs. gate-source voltage.

symbol shown in Fig. 1-5, note that the direction of the arrow at the
gate-channel junction shows the direction of easy gate-current flow. (This
is similar to the *p-n* diode symbol.) In its reverse-bias mode, the gate
"leakage" current may be only a few picoamperes; thus the low-frequency
input impedance of the FET is very high. In most JFET applications
the forward gate-bias mode is not used because of the resulting low

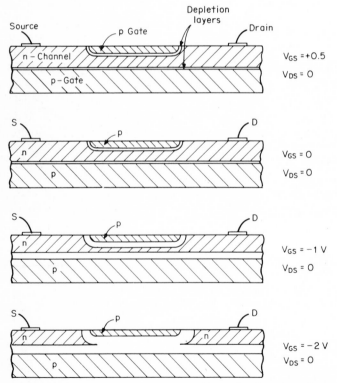

**FIGURE 1-7   Concept of depletion in channel as function of gate-
to-channel voltage ($V_{DS} = 0$).**

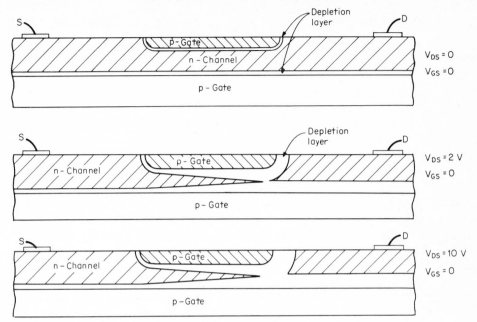

FIGURE 1-8   Channel depletion vs. drain-source voltage.

input impedance. At some relatively high reverse gate-channel voltage, avalanche breakdown will occur at the *p-n* junction. This places an upper limit on the device operating voltage.

Figure 1-7 presents a concept of the depletion in the channel at various gate-channel voltages. These sketches assume zero drain current so that there is no voltage drop along the channel length. If a positive voltage is applied to the drain, with the gate-source voltage at zero, a nonuniform depletion width occurs along the channel as shown in Fig. 1-8. As V_{DS} is made more positive, the depletion layer at the drain end of the channel increases, and thus the incremental channel conduction decreases; if the channel is long (compared to its thickness), then the channel will "pinch off" at the drain end and channel current will saturate at a drain-gate voltage approximately equal to $-V_{GS(off)}$. Further increases in drain voltage will have little effect upon drain current; hence the current "saturates."

Figure 1-9 shows the drain current I_D vs. drain-source voltage V_{DS} with gate-source voltage V_{GS} at zero. As V_{DS} is made more positive, the depletion layer at the drain end of the channel increases, and thus the incremental channel conduction decreases. The slope $\Delta I_D/\Delta V_{DS}$ continues to decrease as V_{DS} is increased, until at some high voltage breakdown occurs. In the negative quadrant a rapid increase in drain current occurs because the drain-gate junction becomes forward-biased. The

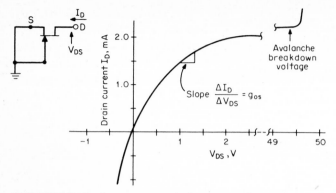

FIGURE 1-9   Drain current vs. drain-source voltage.

JFET is not normally operated in this quadrant more than a few tenths of a volt. Figure 1-10 shows the magnitude of $\Delta I_D/\Delta V_{DS}$ as a function of V_{DS} for a typical FET. This is the output conductance g_{os} of the FET operating in a common-source configuration.

With reference to Fig. 1-8, it may appear that when V_{DS} exceeds "pinch-off," no current can flow. However, there does exist an electric field across this "depletion" region and a supply of carriers (electrons) at the source end of the channel; thus carriers will drift across the depletion region just as they drift across the collector-base depletion region in a bipolar transistor. As V_{DS}, approaches the pinchoff value, the drain current I_D tends to limit at a saturation level.

1-2-2   Effect of Channel Length

The degree of saturation is a function of the device geometry. A long channel will become more saturated (have a lower $\Delta I_D/\Delta V_{DS}$) than a

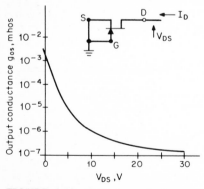

FIGURE 1-10   Output conductance vs. drain-source voltage.

short channel; however, the long channel will also have a lower initial channel conductance than will the short channel. For the short-channel device (where length-to-thickness ratio is less than about 4), drain-voltage "pinchoff" may be higher than $-V_{GS(off)}$. This is because the minimum channel width does not occur directly under the gate junction but is shifted toward the drain terminal. A concept of the shape of a short channel vs. drain voltage is shown in Fig. 1-11. It has been calculated[1] that for a channel length-to-thickness ratio of unity, drain voltage at which saturation occurs will be approximately 1.6 times $-V_{GS(off)}$.

For the short-channel FET an effect known as carrier "velocity saturation" may cause drain current to saturate before minimum channel width is reached. This effect occurs because at high electric fields, carrier mobil-

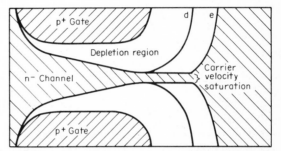

FIGURE 1-11   Shape of conductive channel when carrier velocity saturation occurs. (Reprinted with permission from *Solid State Electronics*, vol. 10, J. R. Hauser, "Characteristics of JFET Devices with Small Channel Length-to-Width Ratios," © 1967, Pergamon Press, Ltd.)

ity is no longer constant but becomes inversely proportional to the electric field. For fields greater than this critical value, drift velocity no longer increases with increasing field; hence the drain current saturates. Figure 1-11 shows channel shape, calculated by Hauser,[1] when carrier velocity saturation occurs. Most devices designed for high-frequency and high-power application have short channels. For these devices drain-current saturation may be due to carrier velocity saturation instead of channel "pinchoff."

1-2-3   Transconductance (g_{fs})

A useful circuit design characteristic is the effect of gate-source voltage upon drain current. A family of these I_D versus V_{GS} "transconductance" characteristics for a long-channel FET is shown in Fig. 1-12. The data

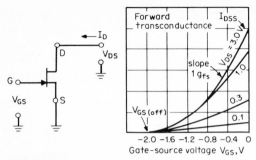

FIGURE 1-12   Forward transfer characteristic of
n-channel JFET (slope $\Delta I_D / \Delta V_{GS} = g_{fs}$).

for these curves were obtained using a 2N4868 which had a $V_{GS(off)}$ of approximately 2 V. If V_{DS} is greater than $-V_{GS(off)}$, then the shape of the transconductance curve is fairly independent of V_{DS}. The slope $(\Delta I_{DS}/ \Delta V_{GS})$ is the forward transconductance g_{fs}. Figure 1-13 shows g_{fs} vs. V_{GS} for $V_{DS} \geqslant -V_{GS(off)}$. The similarity of the curve of Fig. 1-13 to Fig. 1-5 is normal; g_{fs} and g_{os} are approximately linear functions of $(V_{GS} - V_{GS(off)})$.

Another characteristic of interest to circuit designers is the relationship g_{fs} vs. I_D. Such a curve is shown in Fig. 1-14. This curve indicates that g_{fs} is an approximate function of $(I_D)^{1/2}$. This square-root relationship is common for long-channel JFETs and means that in an *RC*-coupled amplifier, voltage gain may be increased by operating at lower drain currents, if $I_D R_L$ is kept constant.

A family of common-source output characteristics is given in Fig. 1-15. Note that as the gate is made more negative, the saturated value of I_D is decreased and the value of V_{DS} at which saturation occurs is decreased by the approximate magnitude of V_{GS}. Drain-current saturation

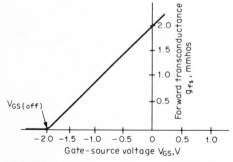

FIGURE 1-13   Forward transconductance vs.
gate-source voltage.

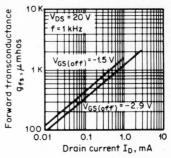

FIGURE 1-14 Forward transconductance vs. drain current.

occurs at a fixed drain-to-gate voltage; thus, if the gate is biased negative
with reference to the source, saturation occurs at a lower drain-source
voltage.

1-2-4   Capacitance

The capacitance per unit area of the p-n junction between the gate and
the source and drain is a function of the junction depletion-layer thickness. The depletion layer acts as a dielectric between the p and n conducting regions. A reverse bias on the p-n junction causes the depletion
layer to increase and thus the capacitance to decrease. Concepts of the
junction depletion as a function of gate-to-channel voltage and of drain-
to-source voltage are shown in Figs. 1-7 and 1-8. Theory indicates that
the capacitance of an abrupt junction is an inverse function of the square
root of the junction voltage. In most amplifier applications the drain-
gate voltage is greater than the source-gate voltage; thus the drain-gate
capacitance will be lower than the source-gate capacitance.

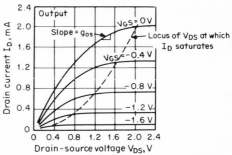

FIGURE 1-15   Common-source output characteristics (2N4868).

1-2-5 Temperature Effects

The effect of device temperature upon device characteristics can be related to three factors:

1. Carrier *mobility* is an inverse function of temperature. This causes the channel conductivity to have a negative temperature coefficient.

2. Ionization of carriers is a function of temperature. For an *n*-channel JFET the donor impurities in the channel region are fully ionized at about $-200°C$; thus in the -55 to $+125°C$ range, the effect of temperature upon channel carrier density is small. However, thermal generation of hole-electron pairs within the depletion region of the gate-channel junction results in a significant reverse gate-current vs. temperature coefficient. This effect results in an approximate doubling of reverse gate "leakage" current for each 10 to 12°C increase in temperature.

3. Temperature increases also cause a decrease in the *thickness* of the depletion layer at the channel-gate junction. This results in an increase in channel thickness and an increase in the magnitude of $V_{GS(off)}$.

Characteristics 1 and 3 have opposing effects upon channel conduction. This is illustrated in Fig. 1-16, which shows transfer characteristics at three temperatures. For this device, I_D at $V_{GS} = 0$ has a negative temperature coefficient. This is due to the negative temperature coeffi-

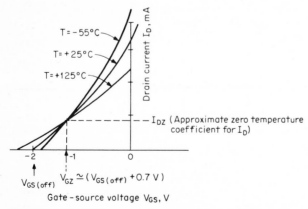

FIGURE 1-16   FET transconductance characteristics vs. temperature.

cient of carrier mobility. At a value of V_{GS} about 0.7 V above $V_{GS(off)}$, I_D has a near-zero temperature coefficient because these two characteristics compensate each other.

1-2-6 *p*-Channel JFET

The foregoing discussion of the JFET has assumed that the channel was *n*-type. A *p*-channel FET has similar characteristics, with the major difference being a change in the polarities. Characteristic curves of a *p*-channel device are given in Fig. 1-17. *P*-type carriers (holes) have lower mobility than *n*-type carriers (electrons); thus for devices of similar dimensions and channel carrier density, the *n*-type FET will have a higher channel conductance than the *p*-type device. For a given channel conductance and pinchoff voltage, the *p*-channel FET will typically have higher interelectrode capacitance and higher junction leakage than the *n*-channel device.

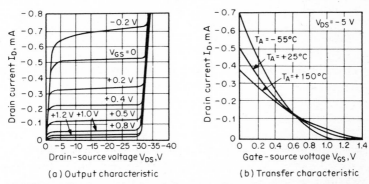

(a) Output characteristic   (b) Transfer characteristic

FIGURE 1-17   Some characteristic curves for *p*-channel JFETs (2N2608).

1-3   THE MOSFET (INSULATED-GATE FET)

A cross section of an *n*-channel MOSFET is shown in Fig. 1-18. With this structure the *n*-channel conduction may be controlled by a voltage applied between gate and source, by one applied between body and source, or by a combination. A very important characteristic of this structure is that the gate is separated from the channel by a very low-leakage dielectric (typically silicon oxide). This permits the gate-channel voltage to be positive or negative. For the *n*-channel structure of Fig. 1-18, a positive gate-source voltage will increase channel conduction, while a negative voltage will decrease channel conduction, as shown in Fig. 1-19. Since no conducting path exists between the control gate and the rest of the structure, the gate-to-channel resistance of a typical device

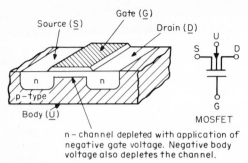

n – channel depleted with application of
negative gate voltage. Negative body
voltage also depletes the channel.

FIGURE 1-18   Cross section of *n*-channel MOS-FET.

(2N3631) exceeds 10^{16} Ω. This is true with positive or negative voltage applied to the gate. An upper limit on gate voltage is imposed by the dielectric breakdown of the thin oxide under the gate metal. Unlike in a *p-n* junction, the breakdown of the oxide dielectric results in permanent damage to the oxide; thus this condition must be avoided.

A family of output characteristics of an *n*-channel MOSFET is given in Fig. 1-20. Note the similarity of these characteristics to those of the *n*-channel JFET shown in Fig. 1-15. The MOSFET characteristics, however, include curves with both positive and negative gate voltage. For the *n*-channel JFET a positive gate voltage would forward-bias the gate-source *p-n* junction. This does not occur with the MOS structure shown. (Some MOSFET devices, however, have a zener diode connection between the gate and the body to limit the voltage that may be impressed across the gate oxide, as a protection for the oxide.) For the *n*-channel MOSFET a positive gate voltage is said to "enhance" the channel; a

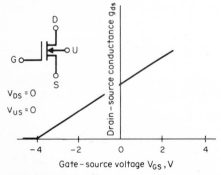

FIGURE 1-19 Channel conduction vs. gate voltage for *n*-channel MOSFET.

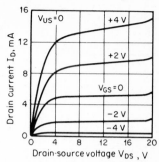

FIGURE 1-20 Common-source output characteristics of n-channel MOSFET (normally ON device.).

negative voltage "depletes" the channel. This is a "normally ON" device—that is, when $V_{GS} = 0$, a conducting channel exists between drain and source.

A normally OFF MOSFET is shown in Fig. 1-21. In this structure no channel exists between drain and source unless a gate voltage is applied. A positive voltage applied to the gate, with respect to source and drain, creates a conducting channel by pulling carriers (electrons for the n-channel type, holes for the p-channel type) from the source and drain regions and from the body into the upper layers of the substrate under the gate. The thickness and conduction of the channel thus created are a function of the gate-source, gate-drain, and gate-body voltages. The gate-source voltage V_{GS} at which channel conduction just begins to occur is called the "threshold voltage" $V_{GS(th)}$. $V_{GS(th)}$ is a function of the thickness of the oxide under the gate, p-type carrier density in the body under the gate, and body-source voltage V_{US}. The body could

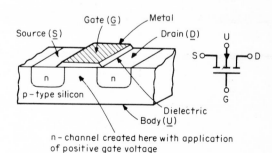

FIGURE 1-21 n-channel MOSFET—normally OFF enhancement type.

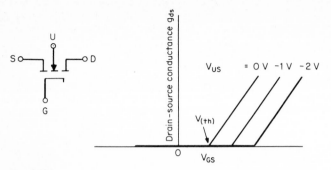

FIGURE 1-22   Channel conductance vs. gate-source voltage for MOSFET showing the effect of "body" bias.

be designed to function as a junction-type control gate but is usually designed to have minimum effect upon the channel conduction and threshold voltage. In many device types the body is internally connected (shorted) to the source terminal, in which case the source-drain voltage is limited to one polarity to avoid forward-biasing the drain-body diode. Figure 1-22 shows channel conduction g_{ds} versus V_{GS} with V_{US} as a parameter. With this particular device, V_{US} has a great effect upon $V_{GS(th)}$ and g_{ds}. In many device designs the body effect is less than that shown.

1-3-1   Dual-Gate MOSFET

A useful structure for high-frequency applications is the dual-gate MOSFET. This device, as shown in Fig. 1-23, has two control gates between the drain and the source. Typically in a high-frequency amplifier, G_1 is utilized as the signal input terminal and G_2 is at signal ground. This results in very low drain-to-G_1 feedback capacitance. This operating mode is similar to the "Cascode" operation of vacuum tubes. G_2 is also commonly utilized as an automatic gain control element.

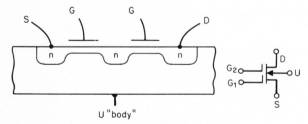

FIGURE 1-23   Dual-gate MOSFET.

1-4   FET SYMBOLS

The FET "family tree" shown in Fig. 1-2 indicates six different types. A different schematic symbol is utilized for each of these types. Figure 1-24 relates the symbols to the device types. Note that the direction of the arrow between the gate or body terminal and the channel will indicate the channel type. The "channel" of the enhancement types of MOS is broken, indicating that a channel must be "enhanced" to make the device operative.

These symbols will be utilized in the chapters that follow. Understanding these symbols will aid in understanding symbols for new device types that may be developed. Many device types are symmetrical with respect to source and drain; that is, source and drain functions may be interchanged. However, there are also devices which are not symmetrical. The symbol locates the gate electrode adjacent to the end of the device which is characterized by the manufacturer as the "source"

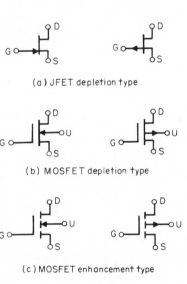

(a) JFET depletion type

(b) MOSFET depletion type

(c) MOSFET enhancement type

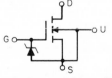

(d) MOSFET enhancement type with internal zener diode to protect gate oxide

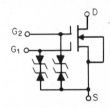

(e) n-channel dual-gate depletion type MOSFET with gate protection diodes and body internally connected to source

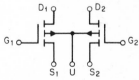

(f) Dual p-channel enhancement-type MOSFET with common body

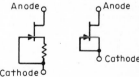

(g) n-channel JFETs designed to function as a "constant current" diode

FIGURE 1-24   FET symbols.

terminal. (An attempt has been made to utilize symbols that correspond to IEEE/ANSI/IEC and industry standards.)

As a word of caution, it should be pointed out that many schematics in technical literature neglect to show the "body" terminal of MOSFETs. In *most* cases it may be assumed that the body is internally connected to the source terminal; however, there are many cases in which the body is connected to some other potential, especially when the device is used as an analog switch. In integrated circuits, the body is often common to many MOSFETs within the same circuit.

1-5   PHYSICAL CHARACTERISTICS OF FETS

FET design and manufacturing details are beyond the scope of this book; however, a brief survey of some of the physical characteristics may be worthwhile. Manufacturing methods are similar to those utilized in other semiconductor devices such as integrated circuits, transistors, and diodes. Dimensions of the active areas, such as channel thickness, width, and length, may be controlled by a combination of epitaxial thickness, diffusion, ion implanting, etching, and photolithographic techniques.

1-5-1   Typical JFET Structures

Figure 1-25 shows some of the steps involved in the manufacture of one type of *n*-channel JFET. The starting material in this case is a wafer of high-purity monocrystalline silicon which has been doped with an acceptor impurity (*p*-type). Onto this wafer a layer of silicon is epitaxially grown to a precisely controlled thickness and with a closely controlled donor (*n*-type) impurity concentration. A layer of silicon dioxide (glass) is formed on the wafer surface. By photolithographic techniques certain regions of this oxide are removed. The remaining oxide functions as a mask or barrier to *p*-type atoms, which are diffused into the unmasked regions. The masked region shown in Fig. 1-25*c* will become the source, drain, and channel region of the finished FET. A second oxidation, masking, and diffusion operation is used to create a top *p*-type gate which separates the source and drain. The depth of the top-gate diffusion and the initial thickness of the epitaxial layers determine the channel thickness. The width of the top gate stripe determines the channel length (source-to-drain dimension). Small variations in photographic dimensions, diffusion depths, epitaxial-layer thickness, and donor and acceptor impurity density will have an effect upon channel conduction, pinchoff voltage, gate-to-channel capacitance, and junction breakdown voltage. Lack of precise control of the physical parameters results in a production

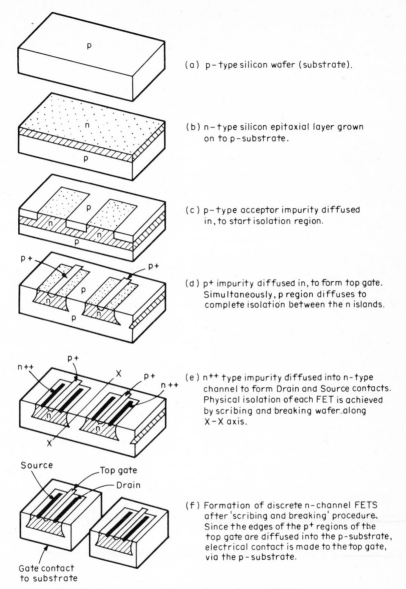

(a) p–type silicon wafer (substrate).

(b) n–type silicon epitaxial layer grown on to p–substrate.

(c) p–type acceptor impurity diffused in, to start isolation region.

(d) p+ impurity diffused in, to form top gate. Simultaneously, p region diffuses to complete isolation between the n islands.

(e) n++ type impurity diffused into n–type channel to form Drain and Source contacts. Physical isolation of each FET is achieved by scribing and breaking wafer along X–X axis.

(f) Formation of discrete n–channel FETS after 'scribing and breaking' procedure. Since the edges of the p+ regions of the top gate are diffused into the p–substrate, electrical contact is made to the top gate, via the p–substrate.

FIGURE 1-25 Some assembly steps for *n*-channel JFETs.

spread of the electrical characteristics of a given device type. In many instances a family of device types may be produced from a given set of photographic masks by varying the thickness of the epitaxial layer and the depth of the top-gate diffusion.

For a given channel thickness, width, and carrier concentrations, conduction will be an inverse function of channel length. For the structure described in Fig. 1-25 this length is controlled by the photographic process of etching the oxide window for top-gate diffusion, and by the sideways diffusion of the top gate. As the window size is decreased to make shorter, higher-conduction channels, control of the photographic process becomes more difficult. Because short channels are important for high-conduction and high-frequency devices, several methods for creating short-channel structures not so dependent upon very small photographic dimensions have been developed.

1-5-2   Short-Channel MOSFETs (VMOS and DMOS)

Two significant techniques are being utilized to produce short-channel MOSFETs. Both utilize the difference in depth of two diffusions instead of photolithographic techniques to control channel length. The differences in the two techniques relate to the geometry employed. In one type the channel is oriented parallel to the chip surface (as with the previously described MOSFET). It is commonly called the DMOS (for diffused-channel MOS). The second type has its channel oriented approximately vertical to the chip surface. While its channel is also controlled by diffusion, it is commonly called the VMOS (for vertical-channel MOS). Section views of the two types are shown in Fig. 1-26, both of which are n-channel types. The p-type "body" diffusions and the n^+ type source diffusions in the DMOS may be made through the same window opening in the oxide. The p type is allowed to diffuse slightly deeper than the n type; the difference becomes the channel length. The n^+ drain contact for the DMOS is diffused through a window in the oxide a short distance from the drain end of the channel region. The n^+ drain contact for the VMOS is the substrate of the chip; thus the entire back side forms the drain. In the DMOS the gate metal is deposited on the surface of the oxide above the p^- region where the channel is formed. For the VMOS, a V groove is first etched down through the n^+ source and the p^- body regions. The vertical dimension of this V groove is determined by the width of the oxide window and the crystallographic structure of the silicon wafer. This is achieved by using a chemical that preferentially etches along certain crystal planes. As with the other MOS types described, the process is completed by growing oxide over the channel region and applying metalization to the gate, source, and body regions (for this device the drain is the back of the chip).

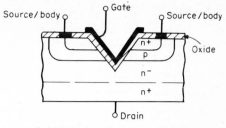

(a) VMOS (vertical channel MOS)

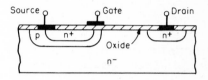

(b) DMOS (double-diffused MOS)

FIGURE 1-26   Short-channel MOS structures.

The n^- region between the drain contact and the "body" forms a depletion region which results in a low drain-to-body and drain-to-gate capacitance—an important feature for high-frequency applications and high-speed switching. The n^- region also provides a means of achieving a high drain-voltage rating while still maintaining a short, high-conductance channel. The VMOS structure gives better utilization of the silicon surface because the substrate is utilized for the drain. Also, both sides of the V groove are utilized for channels; thus two current paths are available for a single gate finger.

The short-channel VMOS structure has permitted the introduction of MOSFETs into high-power, high-frequency applications. No minority-carrier storage times (as encountered with bipolar types) are involved in switching the unit OFF; switching 1 A of drain current in less than 4 ns is typical. Figure 1-27 shows some of the characteristics of Siliconix VMOS type 2N6661. This unit has a 2-A, 90-V rating. The drain current has a negative temperature coefficient; this means that in power applications devices can be paralleled without encountering "current hogging" problems of the type encountered when bipolar transistors are paralleled.

1-5-3   Short-Channel JFET

The short-channel JFET has a high conductance advantage, similar to the short-channel MOSFET. Vertical channel structures have been described by Teszner and Gicquel,[3] by R. Zuleeg,[4] and by others. Because

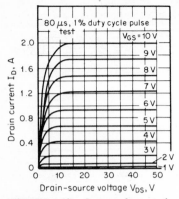

FIGURE 1-27 **Output characteristics of VMOS (Siliconix type 2N6661).**

of their relatively high input capacitance, their use has been primarily limited to linear hi-fi equipment.

A section view of a vertical short-channel JFET as described by Zuleeg[4] is shown in Fig. 1-28. Here the *p*-type gate structure is buried within the *n*-channel region separating the source and the drain.

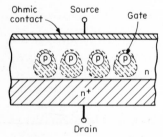

FIGURE 1-28 Gate structure and depletion region shape near channel pinchoff in a vertical-channel JFET. (Reprinted with permission from *Solid State Electronics*, vol. 10, R. Zuleeg, "Multi-channel Field-Effect Transistor, Theory and Experiment," © 1967, Pergamon Press, Ltd.)

1-5-4   Gallium Arsenide FETs (GaAs FET)

The majority of the FETs currently available utilize silicon as the semi-conductor material. Additional materials are being used in the development of new FETs, primarily aimed at achieving higher-gain bandwidths and lower high-frequency noise. The ideal high-frequency FET needs a base material that can simultaneously offer high carrier mobility, high drift velocity, and high avalanche breakdown. The semiconductor crystal gallium arsenide (GaAs) offers mobility seven times that of silicon and drift velocity 70 percent over silicon. However, the avalanche breakdown field for GaAs is only one-fourth that of silicon; thus high-power performance requires much higher currents. Because the techniques are less advanced, manufacturing costs are greater for GaAs than for silicon. These devices are making a significant contribution to amplifier design in the gigahertz range.

1-6   SUMMARY

This chapter has given a qualitative description of several types of field-effect transistors. As the presentation is aimed at the FET user, device design details have been avoided.

Chapter 2 presents a more detailed electrical description of the FET, including mathematical interrelationships of the various characteristics such as pinchoff voltage, drain saturation current, transconductance, channel ON resistance, temperature coefficients, and inner-electrode capacitance.

Other chapters will be devoted to applications of FETs in such circuit functions as analog switches, amplifiers, mixers, oscillators, voltage converters, voltage-controlled resistors, and constant current sources.

Although the commercial availability of FETs followed the bipolar transistor by 8 to 10 years, this device has become a very valuable component in electronic circuits. Integrated circuits in such devices as computers, memories, and electronic TV games make extensive use of MOSFETs. TV and FM tuners and hi-fi amplifiers utilize FETs as input stages because of their superior low-noise performance. Ionization-chamber-type smoke detectors utilize a MOSFET input-stage amplifier because of the very low leakage of the MOSFET gate. Analog multiplexer systems incorporate FETs and analog switches because of the absence of junction offsets in the ON state and the bidirectional blocking capability in the OFF state. Recent developments in the manufacture of short-channel FETs, such as VMOS, have increased power capabilities and have resulted in increasing high-power, high-frequency applications, a few of which are presented in the chapters that follow.

REFERENCES

1. Hauser, J. R., "Characteristics of JFET Devices with Small Channel Length-to-Width Ratios," *Solid State Electronics*, **10**:577–587, 1967.

2. Shockley, W., "A Unipolar Field-Effect Transistor," *Proceedings, IRE*, **40**:1365–1376, 1952.

3. Teszner, S., and R. Gicquel, "Gridistor—A New Field-Effect Device," *Proceedings, IEEE*, **52**:1502–1513, 1964.

4. Zuleeg, R., "Multi-channel Field-Effect Transistor, Theory and Experiment," *Solid-State Electronics*, **10**:559–576, 1967.

BIBLIOGRAPHY

Das, M. B., and **J. M. Moore:** "Measurements and Interpretation of Low-Frequency Noise in FETs," *IEEE Transactions on Electron Devices*, **ED-21** (4):247–257, April 1974.

Hartmann, K.: "Noise Characterization of Linear Circuits," *IEEE Transactions on Circuits and Systems*, **10**:581–589, October 1976. "Japanese Take Two Steps Forward in MOS-Bipolar Compatibility," *Electronics*, p. 207, October 13, 1969.

Sam, C. T.: "Theory of Low-Frequency Generation Noise in JFETs," *Proceedings, IEEE*, **52**:795–814, 1964.

Sze, S. M.: *Physics of Semiconductor Devices*, John Wiley and Sons, Inc., New York, 1969.

Vander Kooi, M., and **L. Ragle:** "MOS Moves into Higher-Power Applications," *Electronics*, 98–103, June 24, 1976.

van der Ziel, A.: "Gate Noise in FETs at Moderately High Frequencies," *Proceedings, IEEE*, **51**:461–467, 1963.

——: "Thermal Noise in FETs," *Proceedings, IRE*, **50**:1808–1812, 1962.

2

PARAMETERS AND
SPECIFICATIONS

2-1 Introduction
2-2 Static Characteristics, JFET
2-3 Gate-Current Characteristics
2-4 Small-Signal Coefficients
2-5 JFET Capacitances
2-6 Noise Characteristics
2-7 Characteristic Interrelationships
2-8 MOSFET Characteristics
2-9 Analog Switch Parameters
2-10 Temperature Effects
2-11 Derating Factors
2-12 Glossary of Terms and Abbreviations

2-1 INTRODUCTION

Chapter 1 was intended primarily to impart an understanding of the underlying phenomena of the field-effect transistor rather than to serve as an exact analysis of its behavior. To derive the relationship between I_D, V_{DS}, and V_{GS} from purely theoretical considerations would be difficult and is hardly justifiable from a user's standpoint. From a practical standpoint, it is often preferable to represent the device characteristics by experimentally determined curves rather than by exact mathematical expressions, because the effects of donor and acceptor impurity concentrations, initial channel and depletion-layer thickness, photographic mask configurations, and alignments, and variations of these within a device type are thus taken into account. The equations presented in this chapter should not be considered exact, but do give close approximations of the performance of most devices. They follow, for the most part, FET

25

theory presented by Shockley[5] and others[1,2]; however, certain physical characteristics assumed in theory are sometimes difficult to achieve in practice. For example, the original Shockley theory assumed abrupt junctions, uniform channel conductivity, and a carrier mobility independent of electric field intensity. Real devices, made by diffusion techniques, have some gradation of junctions and channel conductivity. In short-channel FETs the field at the drain end of the channel is typically high enough to cause some reduction in carrier mobility.

For the initial discussions about static characteristics and the interrelationship of various parameters, we will use as an example a device which has a relatively long channel. Later the effect of shortening the channel length will be discussed.

2-2   STATIC CHARACTERISTICS, JFET

In general, in a three-terminal device, the drain current I_D is a function of two variables, V_{DS} and V_{GS}. This function is best represented by families of characteristic curves, as shown in Fig. 2-1. These curves are for the "common-source" configuration, with the drain as the output and the

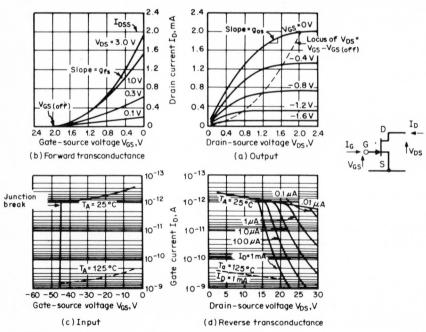

FIGURE 2-1   Static characteristics—n-channel JFET.

gate as the input. They reveal that for this device, if V_{DS} is greater than about 2 V (but less than the drain breakdown), I_D is primarily determined by the gate voltage V_{GS}. Under these conditions it is valid for small signals to characterize the FET by a single transfer characteristic curve, commonly called the "forward transconductance curve," such as the upper curve shown in Fig. 2-1b. Some of the relationships between the forward transfer curve and the output characteristic curves of Fig. 2-1a will be examined. The value of V_{GS} which reduces I_D to approximately zero is the gate-source cutoff voltage, $V_{GS(off)}$. With reference to the output curve, Fig. 2-1a, note that the drain current at $V_{GS} = 0$ tends to become saturated at a drain voltage approximately equal in magnitude to $-V_{GS(off)}$. This drain voltage is often referred to as the pinchoff voltage V_p; however, in this text pinchoff voltage is used interchangeably with gate-source cutoff voltage. V_P will have the same meaning as $V_{GS(off)}$. The symbol I_{DSS} is commonly used to indicate the value of saturated drain current at $V_{GS} = 0$.

The forward transconductance characteristic of Fig. 2-1b can be approximated by a power law relation expressed as

$$I_D = I_{DSS} \left(1 - \frac{V_{GS}}{V_{GS(off)}} \right)^n \tag{2-1}$$

if $V_{DS} \geqslant -V_{GS(off)}$. By differentiation the small-signal transconductance g_{fs} is given by

$$g_{fs} = \frac{dI_D}{dV_{GS}} = -n \frac{I_{DSS}}{V_{GS(off)}} \left(1 - \frac{V_{GS}}{V_{GS(off)}} \right)^{n-1} \tag{2-2}$$

Some texts indicate a value of 3/2 for n; however, experimental measurements on a number of n-channel JFET geometries indicate that the exponent n is close to 2, which is the value derived in an approximate treatment by R. D. Middlebrook.[2]

A useful relationship between g_{fso}, I_{DSS}, and $V_{GS(off)}$ is derived from the ratio of Eqs. (2-2) and (2-1).

$$\frac{g_{fs}}{I_D} = n(V_{GS} - V_{GS(off)})^{-1} \tag{2-3}$$

At $V_{GS} = 0$, $I_D = I_{DSS}$ and $g_{fs} = g_{fso}$. Using 2 as the value of the constant n leads to

$$g_{fso} = -2 \frac{I_{DSS}}{V_{GS(off)}} \tag{2-4}$$

For n-channel FETs I_{DSS} is positive and $V_{GS(off)}$ is negative; for p-channel FETs I_{DSS} is negative and $V_{GS(off)}$ is positive; thus g_{fs} is a positive quantity for both p- and n-channel FETs.

Equation (2-1) indicates that for $V_{GS} = V_{GS(off)}$, $I_D = 0$. In a real device

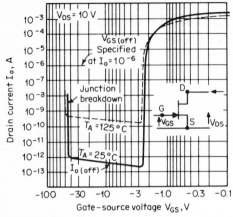

FIGURE 2-2   Drain current vs. gate-source voltage—*n*-channel JFET (2N4868A).

this does not happen. Starting from zero, as the gate voltage is made more negative, the drain current decreases until it reaches a very low value equal to the drain-gate leakage current. At this value the source current will consist of source-gate leakage. Any further increase in the magnitude of the negative gate voltage will result in an increase in I_D leakage. For the small-signal device illustrated, this minimum I_D is on the order of 2×10^{-13} A.

A more detailed I_D-vs.-V_{GS} characteristic is shown in Fig. 2-2, in which the data are plotted on logarithmic paper to show the current magnitude near "cutoff." Because I_D does not go to zero, the error of Eq. (2-1) increases as V_{GS} approaches $V_{GS(off)}$. From a practical measurement standpoint, $V_{GS(off)}$ is usually specified at an I_D value greater than the minimum expected value. The symbol $I_{D(off)}$ is used for the approximate minimum value of I_D. Device types characterized for switching applications will usually specify both $V_{GS(off)}$ and $I_{D(off)}$, as shown in the example for the 2N3970 specification:

Characteristic	Min	Max	Unit	Text conditions
$I_{D(off)}$ drain cutoff current		250	pA	$V_{DS} = 20$ V, $V_{GS} = -12$ V
$V_{GS(off)}$ gate-source cutoff voltage	-4	-10	V	$V_{DS} = 20$ V, $I_D = 1$ nA

For this device $V_{GS(off)}$ is specified as that value of V_{GS} required to reduce I_D to 1 nA, with a V_{DS} of 20 V.

As is shown by the data plotted in Fig. 2-2, when the FET is to be used as a switch, the value of V_{GS} for the switch OFF condition should be a little greater than the data sheet–specified maximum $V_{GS(off)}$.

Figure 2-2 shows that at a very high negative value of V_{GS}, there occurs a rapid increase in I_D. This is the voltage at which avalanche "breakdown" occurs at the drain-gate n-p junction; it sets an absolute maximum rating on the drain-gate voltage for the device. In practice, the device data sheet will usually specify a minimum value for gate-to-drain and gate-to-source breakdown, as shown in the 2N3970 specification:

Characteristic	Min	Max	Unit	Test conditions
BV_{GSS} gate reverse breakdown voltage	-40		V	$I_G = 1\ \mu A,\ V_{DS} = 0$

Since the test conditions set $V_{DS} = 0$, this specification ensures that both drain-to-gate and source-to-gate breakdown will be equal to or greater than 40 V. The value of I_G for the test condition is greater than the normal gate leakage current of the device. I_G-vs.-V_{GS} curves are shown in Fig. 2-3 with test points for BV_{GSS} and I_{GSS} indicated. I_{GSS} is the gate leakage current at a specified value of V_{GS} with $V_{DS} = 0$. Forward gate characteristics are not usually specified because few applications require operation in this mode. The curve of Fig. 2-3b is typical

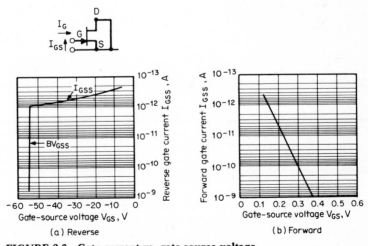

(a) Reverse (b) Forward

FIGURE 2-3 Gate current vs. gate-source voltage.

for a small-signal JFET with the gate *forward*-biased. Operation with a few tenths of a volt forward gate signal is satisfactory if a few hundred picoamperes of gate current permits proper circuit performance.

The output characteristic curve for $V_{GS} = 0$ of Fig. 2-1a approximately follows the equation

$$I_D = g_{dso} V_{DS} \left(1 + \frac{V_{DS}}{2V_{GS(off)}} \right) \tag{2-5}$$

where g_{dso} = slope dI_D/dV_{DS} at V_{GS} and $V_{DS} = 0$. It has been shown that $g_{dso} = g_{fso}$; therefore, by utilizing Eq. (2-4), Eq. (2-5) can be given as

$$I_D = \frac{2I_{DSS}}{-V_{GS(off)}} V_{DS} \left(1 + \frac{V_{DS}}{2V_{GS(off)}} \right) \tag{2-6}$$

If we rearrange terms and for convenience let $V_p = V_{GS(off)}$, we get

$$I_D = I_{DSS} \left[2\frac{V_{DS}}{-V_p} - \left(\frac{V_{DS}}{V_p} \right)^2 \right] \tag{2-7}$$

This equation is not valid when V_{DS} exceeds $-V_p$. It indicates that as V_{DS} approaches $-V_p$, the rate of change of I_D with changing V_{DS} decreases and reaches zero at $V_{DS} = V_p$. In a real device dI_D/dV_{DS} does not reach zero.

For some types of applications of the FET, it is helpful to understand the characteristics at very low values of V_{DS}, such as those shown in Fig. 2-4. A "very low value" is one which is small compared to the magnitude of $V_{GS} - V_{GS(off)}$. In this region V_{DS} is small enough to have little effect upon channel thickness, so that the I_D/V_{DS} slope is nearly linear. Since the slope is a function of V_{GS}, the FET can be utilized as a voltage-controlled resistor. The conductance slope ($\Delta I_D/V_{DS}$) at $V_{DS} = 0$ is approximately a linear function of $V_{GS} - V_{GS(off)}$.

If g_{ds} at $V_{GS} = 0$ is given the term g_{dso}, then

$$g_{ds} = g_{dso} \left(1 - \frac{V_{GS}}{V_{GS(off)}} \right) \tag{2-8}$$

with $V_{DS} = 0$. A plot of this characteristic is shown in Fig. 2-5, along with g_{fs} and I_D characteristics.

The relationship between g_{dso}, $V_{GS(off)}$, and I_{DSS} is given by the equation

$$g_{dso} = -2\frac{I_{DSS}}{V_{GS(off)}} \tag{2-9}$$

where I_{DSS} and $V_{GS(off)}$ are as indicated in Fig. 2-1. It is important to note that Eqs. (2-8) and (2-9) and the g_{ds} curve of Fig. 2-5 are valid

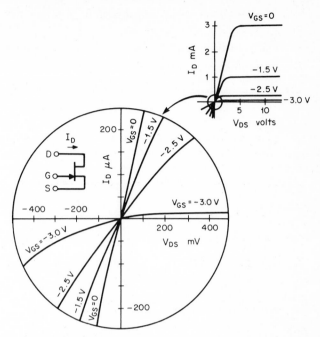

FIGURE 2-4 *n*-channel JFET output characteristic enlarged around $V_{DS} = 0$.

only for the case where V_{DS} is very small compared to V_p. Drain-source conductance at higher values of V_{DS} will be discussed later. FETs designed to be used as voltage-controlled resistors typically have a high $V_{GS(off)}$ because the $V_{DS}/(V_{GS} - V_{GS(off)})$ ratio should be low to keep distortion low.

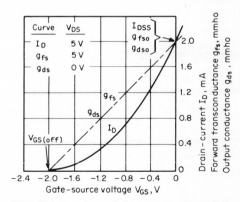

FIGURE 2-5 I_D, g_{fs}, and g_{ds} versus gate-source voltage.

In analog switching applications, V_{GS} is usually set at zero for the ON condition and at a voltage greater than $V_{GS(off)}$ for the OFF condition. A maximum value of ON resistance is usually specified at a small value of I_D or V_{DS}, or by a small-signal ac measurement at $V_{DS} = 0$.

| | 2N3970 | | | |
Characteristic	Min	Max	Unit	Test conditions
$r_{DS(on)}$ static drain-source ON resistance		30	Ω	$V_{GS} = 0$, $I_D = 1$ mA
$r_{ds(on)}$ drain-source ON resistance		30	Ω	$V_{GS} = 0$, $f = 1$ kHz

For the ac method, the signal frequency should be low enough that reactance effects can be neglected.

2-3   GATE-CURRENT CHARACTERISTICS

The gate current I_G of a JFET not only is a function of V_{GS} and V_{GD} but also may be a function of I_D. Figure 2-1d shows gate-current characteristics of type 2N4868A. This device has a BV_{GSS} in excess of 40 V; however, when the device is biased such that drain-to-source current is permitted to flow (a normal amplifier condition), then the gate-current "breakpoint" occurs at a lower drain-to-gate voltage. This breakpoint is a function of basic device design and is usually in the 10- to 30-V range. Beyond this breakpoint, I_G is approximately a *linear* function of I_D and an exponential function of V_{DS}. At $V_{DS} = 0$, the gate current is due to the collection by the gate of thermally generated minority carriers within the space-charge region at the gate-channel junction. When source-to-drain current is permitted to flow through the space-charge region at the drain end of the channel, additional minority carriers are generated due to electrons colliding with silicon atoms. The number of minority carriers created is a linear function of the channel current flowing and an exponential function of the electric field resulting from V_{DG}.

This I_G breakpoint dependency upon I_D occurs at higher voltages in p-channel JFETs because of the lower mobility and lower ionization rates of holes in the drain space-charge region.

For both n-channel and p-channel JFETs the value of I_{GSS} and I_G below the breakpoint is approximately proportional to $\sqrt{V_{GS}}$ and $\sqrt{V_{GD}}$. A further discussion of I_G is given in Sec. 3-8-2.

2-4 SMALL-SIGNAL COEFFICIENTS

Because FETs are employed largely in circuits having varying currents and voltages, their dynamic characteristics are of interest. Referring to Fig. 2-1b, the slope of the curve or rate of change of I_D with respect to V_{GS} is important. This slope is the forward (gate-to-drain) transconductance of the FET, and is given the symbol g_{fs}. As is apparent in Fig. 2-1b, g_{fs} is a function of I_D, decreasing as I_D is decreased. Figure 2-6 shows the relationship between g_{fs} and I_D for a typical FET type. Curves for two devices of the same geometry but having different $V_{GS(off)}$ values are included to show that g_{fs} is also a function of $V_{GS(off)}$. The relationships of these three parameters, shown graphically in Fig. 2-6, are mathematically expressed by Eqs. (2-10), (2-11), and (2-12).

$$g_{fso} = -2\frac{I_{DSS}}{V_{GS(off)}} \tag{2-10}$$

and

$$g_{fs} = g_{fso}\left(1 - \frac{V_{GS}}{V_{GS(off)}}\right) \tag{2-11}$$

also

$$g_{fs} = g_{fso}\left(\frac{I_D}{I_{DSS}}\right)^{1/2} \tag{2-12}$$

The exponent $\frac{1}{2}$ in Eq. (2-12) is indicated by the slope of the curves of Fig. 2-6. A plot of g_{fs} versus V_{GS} is shown in Fig. 2-5. Note the similarity of the curves and equations for g_{fs} and g_{ds}. It should be pointed out that the constant in Eq. (2-10) and the exponent in Eq. (2-12) may not be exact, but for most device designs these numbers will be close.

Referring to Fig. 2-1a, the small-signal output conductance is the slope $\Delta I_D/\Delta V_{DS}$ at a constant value of V_{GS}. The term g_{os} is used to symbolize this parameter. It is apparent that g_{os} is a function of V_{DS} and I_D.

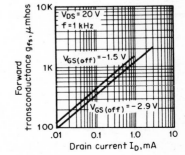

FIGURE 2-6 Common-source forward transconductance vs. drain current.

At $V_{DS} = 0$, g_{os} is equal to the g_{dso} discussed in association with Eq. (2-8). As V_{DS} is increased, g_{os} decreases. For voltage-amplifier applications, a low g_{os} is of importance because the maximum voltage gain is limited by the ratio g_{fs}/g_{os}. If I_D really saturated at $-V_{GS(off)}$, then g_{os} would drop to zero at V_{DG} equal to or greater than $-V_{GS(off)}$. Most device types designed for small-signal low-frequency amplifiers will have g_{fs}/g_{os} ratios in excess of 100 (for $V_{DS} > -V_{GS(off)}$). Figure 2-7 shows g_{os} versus V_{DS} for typical n-channel JFETs. Figure 2-8 shows g_{os} versus I_D for a constant value of V_{DS}. The curves of Figs. 2-6, 2-7, and 2-8 reveal two important points about devices of a given basic design: When operated at a given value of I_D, the device with the lower $V_{GS(off)}$ will have a higher g_{fs} and a lower g_{os}. For a given power-supply voltage, available voltage gain is inversely proportional to $V_{GS(off)}$. Also, lower values of $V_{GS(off)}$ permit use of lower-voltage supplies.

A simplified common-source FET voltage-amplifier circuit and its low-frequency equivalent circuit are shown in Fig. 2-9. This equivalent circuit neglects reactive components and dc leakages. It is given here to show the equivalent circuit location of the g_{fs} and g_{os} parameters discussed above. The voltage amplification with this circuit is

$$A_v = \frac{e_o}{e_g} = -g_{fs}\left(\frac{R_L}{1 + g_{os}R_L}\right) \tag{2-13}$$

If g_{os} is small compared to $1/R_L$, then Eq. (2-8) reduces to

$$A_v = -g_{fs}R_L \tag{2-14}$$

The negative sign indicates that the phase of the signal is inverted.

The low-frequency input conductance g_{iss} contains two principal components: the gate-source conductance ($g_{gs} = \Delta I_G/\Delta V_{GS}$) and the gate-drain conductance ($g_{gd} = \Delta I_G/\Delta V_{GD}$):

$$g_{iss} = g_{gs} + g_{gd} \tag{2-15}$$

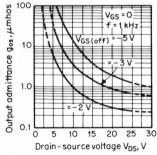

FIGURE 2-7 Output admittance vs. drain-source voltage.

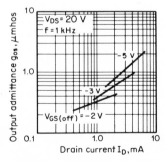

FIGURE 2-8 Output admittance vs. drain current.

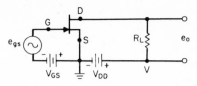

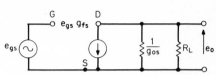

FIGURE 2-9   FET voltage amplifier and equivalent circuit (low-frequency simplified).

As indicated in the curves of Fig. 2-1c and d, g_{iss} for a small-signal device is on the order of 10^{-14} if V_{GS} is below junction breakdown and V_{DS} is below the I_G breakpoint. In an amplifier configuration the effect of g_{gd} will be increased by the voltage gain of the amplifier. For a circuit such as that shown in Fig. 2-9, the low-frequency input conductance would be

$$g_{is} = g_{gs} + g_{gd}(1 + g_{fs}R_L) \tag{2-16}$$

(assuming that $1/g_{os} >> R_L$).

Even with a voltage gain of 10, g_{is} of the FET amplifier will be less

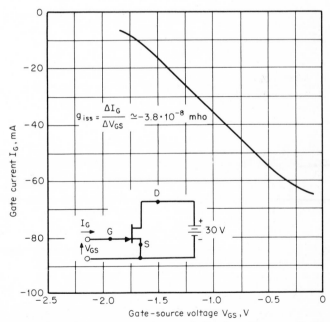

FIGURE 2-10   Gate input characteristic when V_{DS} exceeds the I_G breakpoint.

than 10^{-12} mho at room temperature. Consequently, a FET has a low-frequency input resistance in excess of 10^{12} Ω, since resistance is the reciprocal of conductance.

At values of V_{DS} greater than the I_G breakpoint (see Fig. 2-1d), the dc value of I_G increases sharply. Above this corner the quantity g_{gd} becomes negative, resulting in g_{iss} going through zero and becoming negative. This occurs because I_G in this region is approximately a linear function of I_D. Since $\Delta I_D = g_{fs} \, \Delta V_{GS}$, a positive change in V_{GS} results in an increase in I_D, which in turn causes an increase in the magnitude of I_G. Since I_G is negative, ΔI_G is negative; therefore g_{gd} will be negative. Figure 2-10 is a plot of experimental data obtained from a typical JFET operating with V_{DS} greater than the I_G breakpoint. These data show that the input conductance g_{iss} is negative for this bias condition. A transition from positive g_{iss} to negative g_{iss} will occur near the knee of the I_G-vs.-V_{DS} curves shown in Fig. 2-1d. The FET low-frequency equivalent circuit with the gate conductance components added is shown in Fig. 2-11. The g_{iss} characteristic is not normally specified in FET data sheets; however, its magnitude can be estimated from the I_G specification.

2-5   JFET CAPACITANCES

As the operating frequency is increased, device capacitances become important parameters. For the JFET the principal capacitance is that of the gate-channel junction. The value and geometrical distribution of this capacitance are functions of the voltages V_{GS} and V_{GD} because these voltages have a direct effect upon the junction depletion-layer thickness. For abrupt junctions the capacitance per unit area is

$$C = \left(\frac{K}{V_{bi} + V_G} \right)^{1/2} \tag{2-17}$$

Abrupt junctions are not achieved in the typical FET structure. The channel may consist of an epitaxial n-type layer on a p-type substrate and with a diffused p-type top gate. The diffusion process will cause

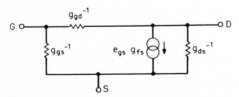

FIGURE 2-11 JFET low-frequency equivalent.

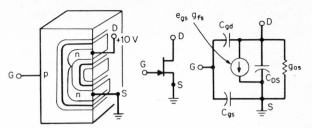

FIGURE 2-12 JFET capacitances, intermediate-frequency model.

some gradation of the junctions. For a graded junction the exponent of Eq. (2-17) would be $\frac{1}{3}$ instead of $\frac{1}{2}$. In a functioning circuit the analysis of gate-to-channel capacitance is complicated by the nonuniform depletion width along the channel. The gate-drain depletion width will be greater than the gate-source depletion width, thus C_{gd} usually will be lower than C_{gs}. Figure 2-12 shows the equivalent-circuit location of these capacitances, plus the drain-source capacitance, and gives a concept of the relative depletion-layer thickness of the drain-gate and source-gate junctions. This sketch is intended to explain why the value of C_{gd} is typically less than that of C_{gs}. In some applications where V_{GD} and V_{GS} are approximately equal (such as in an analog switch or voltage-controlled resistor circuit), C_{gd} and C_{gs} may be approximately equal. The C_{ds} component is largely the device header (package) capacitance and is typically small compared to the other two components; however, at very high frequencies it must be considered. Devices designed specifically for high frequencies often utilize a special low-capacity package configuration.

Figure 2-13 shows the effect of gate voltage upon C_{gs} and C_{gd} for a typical JFET designed for high-frequency (450-MHz) applications (Siliconix type U310). The C_{gd} curve is lower than the C_{gs} because of the 10-V V_{DS} bias (at $V_{GS} = 0$, $V_{GD} = -10$ V). These capacitances increase

FIGURE 2-13 Junction capacitances vs. gate voltage.

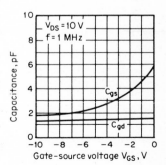

the input and feedback admittance and limit the wideband amplifier performance.

Because FET capacitances are sensitive to applied voltages, the circuit designer must examine the test conditions used in the device data-sheet specifications. For example, the device types listed below can be made from the same basic geometry, but capacitances are specified under different test conditions for each type.

FET type	Characteristic	Max pF	Test conditions	
2N4092	C_{iss} common-source input capacitance	16	$V_{DS} = 20$, $V_{GS} = -5$	$f = 1$ MHz
2N4092	C_{rss} common-source reverse transfer capacitance	5	$V_{DS} = 0$, $V_{GS} = -20$ V	
2N4392	C_{iss}	14	$V_{DR} = 20$, $V_{GS} = 0$	$f = 1$ MHz
2N4392	C_{rss}	3.5	$V_{DS} = 0$, $V_{GS} = -7$ V	
2N4857	C_{iss}	18	$V_{DS} = 0$, $V_{GS} = -10$ V	$f = 1$ MHz
2N4857	C_{rss}	8	$V_{DS} = 0$, $V_{GS} = -10$ V	
2N5564	C_{iss}	12	$V_{DG} = 12$, $I_D = 2$ mA	$f = 1$ MHz
2N5564	C_{rss}	3	$V_{DG} = 12$, $I_D = 2$ mA	$f = 1$ MHz
2N5638	C_{iss}	10	$V_{DS} = 0$, $V_{GS} = -12$ V	$f = 1$ MHz
2N5638	C_{rss}	4		
J111	$C_{dg(\text{off})}$	5	$V_{DS} = 0$, $V_{GS} = -10$	
	$C_{sg(\text{off})}$	5	$V_{DS} = 0$, $V_{GS} = -10$	
	$C_{dg(\text{on})} + C_{sg(\text{on})}$	28	$V_{DS} = V_{GS} = 0$	

2-6 NOISE CHARACTERISTICS

Electrical noise generated within a FET is usually represented by equivalent noise sources, $\bar{e}_n$ and $\bar{i}_n$. Both noise voltage $\bar{e}_n$ and noise current $\bar{i}_n$ are frequency-dependent and have the characteristics shown in Fig. 2-14.

An equivalent noise circuit is shown in Fig. 2-15. Above the frequency f_1, $\bar{e}_n$ is approximately given by:

$$\bar{e}_n \simeq \left(4KTB \frac{0.67}{g_{fs}} \right)^{1/2} \tag{2-18}$$

where $K = 1.374 \times 10^{-23}$ J/K

        $T =$ absolute temperature in kelvins (273 K $= 0°$C)

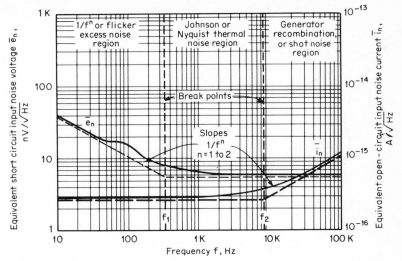

FIGURE 2-14 Characteristics of junction FET noise.

B = frequency range in hertz

g_{fs} = transconductance of FET

With the input short-circuited, the noise voltage across the load R_L resulting from the FET is

$$\text{Output noise voltage} = \bar{e}_n A_V$$

Below the frequency f_1, $\bar{e}_n$ increases proportional to $1/f^n$ and is expressed as

$$\bar{e}_n = \left[4KTB\left(\frac{0.67}{g_{fs}}\right)\left(1 + \frac{f_1}{f^n}\right)\right]^{1/2} \qquad (2\text{-}19)$$

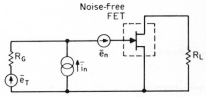

FIGURE 2-15 Equivalent FET noise circuit.

The low-frequency corner f_1 for JFETs is typically in the 100-Hz to 1-kHz range, and the exponent n is usually between 1 and 2. As indicated by the equations, $\bar{e}_n$ is inversely proportional to the square root of g_{fs}.

The equivalent input noise current $\bar{i}_n$ is caused by the current in the gate-to-channel junction. Its approximate value below f_2 is

$$\bar{i}_n = (2qI_GB)^{1/2} \qquad\qquad (2\text{-}20)$$

where $\quad q = 1602 \times 10^{-19}$

$\quad\quad\quad B$ = frequency range in hertz

$\quad\quad\quad I_G$ = dc gate current

This expression is fairly accurate when I_G is the result of the active device conductance. Typically $\bar{i}_n$ will be lower than the calculated value because part of I_G is due to conductance across the device package.

At higher frequencies (above f_2),

$$\bar{i}_n = \left(\frac{4KTB}{R_P}\right)^{1/2} \qquad\qquad (2\text{-}21)$$

where R_P = real part of gate input impedance. The high-frequency corner f_2 is typically in the range of 5 to 50 kHz.

Another form of noise is known as "popcorn" or burst noise, the causes of which have not been completely identified. It shows up as a random short-duration step-function change in drain current, equivalent to an input gate-source voltage change of a few tenths of a microvolt. A more detailed discussion of FET noise is presented in Chapter 3.

2-7 CHARACTERISTIC INTERRELATIONSHIPS

When the time comes to select a FET for a particular circuit application, it is helpful to have an understanding of the interrelationship of the various electrical parameters. If the quantity requirements are to be large, it may be economically desirable to generate a specification tailored to the specific application. If one designs a circuit utilizing a standard catalog device type, for example, and finds that the design can tolerate a wider range of $V_{GS(\text{off})}$ (maximum to minimum), the part supplier may be able to offer a lower-price unit. However, if the range of a related parameter such as the drain saturation current I_{DSS} is not also widened, then the wider $V_{GS(\text{off})}$ spread may be of no help. For a given "geometry" (determined by a set of photographic masks), both $V_{GS(\text{off})}$ and I_{DSS} are affected by conductivity and thickness of the channel. These are determined by one or more non-mask-related process steps, such as epitaxial growth, channel diffusion, gate diffusion, or ion implantation. These parameters are also affected by variations in the masking operations involved in the etching of oxide windows for top gate diffusions, especially on very short channel structures. The approximate equations for the JFET structure indicate the relationship between the electrical parameters and the geometric parameters.

$$g_{fso} = K_1 \frac{W}{L} T \tag{2-22}$$

$$V_{GS(off)} = K_2 T^2 \tag{2-23}$$

$$V_{DSS} = K_3 \frac{W}{L} T^3 \tag{2-24}$$

where W = width
L = length
T = thickness of the channel

If one varies only the thickness, then $V_{GS(off)}$, g_{fso}, and $I_{D(sat)}$ will change in an interrelated manner.

$$I_{DSS} = K_4 (V_{GS(off)})^{3/2} \tag{2-25}$$

$$g_{fso} = K_5 (V_{GS(off)})^{1/2} \tag{2-26}$$

$$g_{fso} = \frac{2 I_{DSS}}{V_{GS(off)}} \tag{2-27}$$

Figure 2-16 shows the interrelationship of these three parameters for a particular JFET geometry. Each dot on these plots represents a particular device. The data were obtained from devices where the only intentional change was in the "top" gate diffusion time to create a change in the channel thickness. The fact that the dots do not all fall on a single line indicates that there were other variations such as donor and acceptor concentrations. This type of variation from the "ideal" control is typical of semiconductor manufacturing processes. From this one geometry or "generic type," several device types may be characterized. For example, the specification limits for three types are indicated on the plot. The complete JEDEC registered characteristics for these device types are given in Table 2-1.

The major differences are due to a difference in channel thickness. The units with a high value of $V_{GS(off)}$ have a thick channel, while those with a low $V_{GS(off)}$ have a thin channel. The data of Fig. 2-16 show that for a given value of $V_{GS(off)}$, the ranges of I_{DSS} and g_{fso} are limited. It should be emphasized that the data are for a particular device geometry and manufacturing process. It is possible to produce devices which meet the specified electrical characteristics using different geometries and processes; therefore, a plot of device types from a different manufacturer will not necessarily coincide with that shown in Fig. 2-16.

The data shown in Fig. 2-16a and b were used to calculate a g_{fso}/I_{DSS} ratio. This characteristic, which is an inverse function of $V_{GS(off)}$, is plotted in Fig. 2-16c. These data indicate that for a given geometry, a higher g_{fso}/I_{DSS} ratio is achieved with the units having lower $V_{GS(off)}$.

TABLE 2-1 ELECTRICAL CHARACTERISTICS (25°C UNLESS OTHERWISE NOTED)*

			2N4867 2N4867A		2N4868 2N4868A		2N4869 2N4869A		Unit	Test conditions
		Characteristic	Min	Max	Min	Max	Min	Max		
1	I_{GSS}	gate reverse current		-0.25		-0.25		-0.25	nA	$V_{GS}=-30$ V, $V_{DS}=0$
2				-0.25		-0.25		-0.25	µA	150°C
3	BV_{GSS}	gate-source breakdown voltage	-40		-40		-40		V	$I_G=-1$ µA, $V_{DS}=0$
4	$V_{GS(off)}$	gate-source cutoff voltage	-0.7	-2	-1	-3	-1.8	-5	V	$V_{DS}=20$ V, $I_D=1$ µA
5	I_{DSS}	saturation drain current (note 2)	0.4	1.2	1	3	2.5	7.5	mA	$V_{DS}=20$ V, $V_{GS}=0$
6	g_{fs}	common-source forward transconductance (note 2)	700	2000	1000	3000	1300	4000	µmho	$V_{DS}=20$ V, $V_{GS}=0$, $f=1$ kHz
7	g_{os}	common-source output conductance		1.5		4		10	µmho	
8	C_{rss}	common-source reverse transfer capacitance		5		5		5	pF	$V_{DS}=20$ V, $V_{GS}=0$, $f=1$ MHz
9	C_{iss}	common-source input capacitance		25		25		25	pF	
10	$\bar{e}_n$	short-circuit equivalent input noise voltage		20		20		20	$\dfrac{nV}{\sqrt{Hz}}$	2N4867 Series, $V_{DS}=10$ V, $V_{GS}=0$, $f=10$ Hz
11				10		10		10		2N4867A Series
12				10		10		10		2N4867 Series, $f=1$ kHz
13				5		5		5		2N4867A Series
14	NF	spot noise figure		1		1		1	dB	$V_{DS}=10$ V, $V_{GS}=0$, $R_{gen}=20$ K, 5 K, 2N4867 Series, 2N4867A Series, $f=1$ kHz

NS

* JEDEC registered data.

NOTES:

[1] Due to symmetrical geometry, these units may be operated with source and drain leads interchanged.

[2] Pulse test duration = 2 ms.

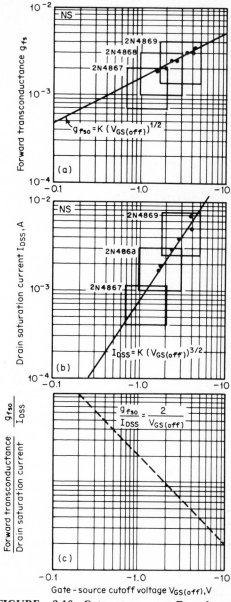

FIGURE 2-16 Gate-source cutoff voltage $V_{GS(off)}$, **in volts, relationship of** I_{DSS}, g_{fso}, **and ratio** g_{fs}/I_{DSS}.

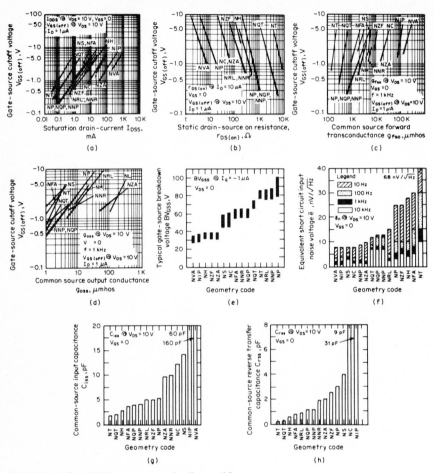

FIGURE 2-17 JFET geometry selection guide.

Characteristics of a variety of *n*-channel JFET geometries are given in Fig. 2-17. The interrelationship of characteristics related to channel thickness is shown in curves *a, b, c,* and *d.* Other geometry-related characteristics are shown in graphs *e, f, g,* and *h.* The geometries are identified by two- or three-letter codes, such as NT, NZA, or NC. These manufacturing codes are used by Siliconix Incorporated to designate their particular FET chip designs and are not necessarily those used by other manufacturers. These letter codes are related to device part numbers in the Geometry section of Siliconix Incorporated's "Field Effect Transistor Data Book." For example, from that book, one can find that the NC geometry is used to make device types 2N3970–72, 2N4091–93, 2N4391–93, 2N4856–61, 2N4856A-6–61A, 2N5564–66, 2N5653–54, and Siliconix type J111–13. As indicated by this list, the NC geometry has wide use. Its relatively short channel gives it a reasonably good g_{fs}/C_{iss} ratio, which makes it useful in UHF/VHF oscillator and amplifier applications up to about 400 MHz.

The NQT geometry is a dual narrow-channel device. It has a low g_{fs}, but because of its small size it has very low gate-leakage current; some matched duals characterized from the NQT geometry have an operating I_G specification of 0.1 pA max.

2-8   MOSFET CHARACTERISTICS

Common-source output and forward transconductance characteristics for two types of MOSFETs are given in Figs. 2-18 and 2-19. Figure 2-18 is for an n-channel depletion-mode device type 2N3631. The output and transfer characteristics indicate that gate-source voltage may be positive or negative.

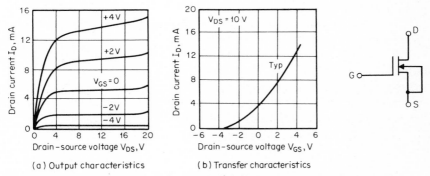

FIGURE 2-18   Characteristics of n-channel depletion-type MOSFET (2N3631).

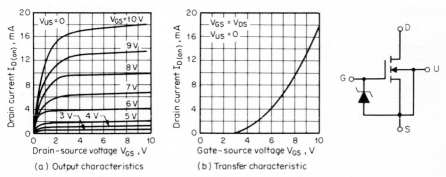

FIGURE 2-19   Characteristics of n-channel enhancement-type MOSFET (M116).

An important characteristic of this device, compared with a JFET, is the very low gate current. Input leakage resistance is typically greater than 10^{15} Ω.

This low input leakage characteristic makes this device useful in ultra-high input-impedance amplifiers for circuits such as ionization-chamber input-type smoke detectors, pH measurement instruments, and proximity sensors. Caution must be exercised in handling this unit because it contains no protective circuit for the gate oxide. A gate-to-channel voltage in excess of ±60 V may cause permanent damage to the oxide. In this particular geometry the body is internally connected to the source.

Figure 2-19 shows characteristics for an n-channel enhancement-mode device which has incorporated a gate-protecting diode connected between the gate terminal and the body of the MOSFET. This diode is designed to have a breakdown voltage of less than the voltage which would damage the gate oxide but greater than the normal gate operating voltage range. The diode limits the magnitude of the negative-going gate-body voltage to about 0.6 V, and the positive gate-body voltage to about 50 V.

The transconductance and output characteristics of the MOSFET are similar to those of the JFET. The transconductance curve is shifted along the V_{GS} axis for the "normally OFF" enhancement-type device, so that a positive gate-source voltage is required to turn this unit ON. The V_{GS} at which channel conductance just starts is called the "gate-source threshold voltage" and is given the symbol $V_{GS(th)}$. This corresponds to the $V_{GS(off)}$ of the depletion-type devices. The transconductance equations of the two types are similar, i.e.,

$$I_D = \frac{K}{2}(V_{GS} - V_{GS(off)})^2 \tag{2-28a}$$

where $V_{GS} > V_{GS(off)}$ (depletion type)

$$I_D = \frac{K}{2}(V_{GS} - V_{GS(th)})^2 \tag{2-28b}$$

where $V_{GS} > V_{GS(th)}$ (enhancement type)
K is the device constant, which is a function of its geometry. The transconductance is

$$g_{fs} = \frac{\Delta I_D}{\Delta V_{GS}} = K(V_{GS} - V_{GS(off)}) \tag{2-29a}$$

where $V_{GS} > V_{GS(off)}$ (depletion type)

$$g_{fs} = \frac{\Delta I_D}{\Delta V_{GS}} = K(V_{GS} - V_{GS(\text{th})}) \tag{2-29b}$$

where $V_{GS} > V_{GS(\text{th})}$ (enhancement type)

Equations (2-28) and (2-29) assume that V_{DS} is greater than $(V_{GS} - V_{GS(\text{off})})$ or $(V_{GS} - V_{GS(\text{th})})$. As with the JFET, for the idealized device, I_D and g_{fs} saturate at a drain voltage of $(V_{GS} - V_{GS(\text{off})})$ or $(V_{GS} - V_{GS(\text{th})})$. Below saturation, I_D and g_{fs} are functions of both V_{GS} and V_{DS}:

$$I_D = K\left(V_P V_{DS} - \frac{V_{DS}^2}{2}\right) \tag{2-30}$$

where $V_P = (V_{GS} - V_{GS(\text{th})})$ or $(V_{GS} - V_{GS(\text{off})})$ and $V_{DS} < V_P$

A low-frequency equivalent circuit is shown in Fig. 2-20 for the common-source configuration. The transconductance g_{fs} has been discussed above. The input conductance g_{in} is due to leakage through the thin gate oxide, device package leakage, and (if included) input protection diode. The output conductance g_{os} has a finite value even at V_{DS} greater than "saturation" because of the decrease in effective channel length with increasing V_{DS}.

The output capacitance C_{ds} consists mostly of the drain-body n-p junction capacitance and is inversely proportional to the square root of the drain voltage. Input capacitance C_{gs} and feedback capacitance C_{dg} are complex functions of V_{GS}, V_{DS}, substrate resistivity, threshold voltage, and overlap of the gate metal above the source and drain regions. Figure 2-21 shows short-circuit gate capacitance vs. gate-source voltage for an n-channel enhancement-type MOSFET. This capacitance is related to the carrier concentration of the silicon directly under the gate metal. At a gate voltage below $V_{GS(\text{th})}$, the "channel" region is depleted of carriers and thus C_g decreases. For V_{GS} more positive than $V_{GS(\text{th})}$, conduction electron concentration is increased; thus C_g increases for $V_{GS} > V_{GS(\text{th})}$. In many devices the metal gate overlaps the drain region. This

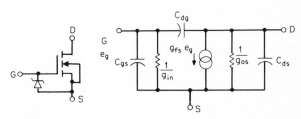

FIGURE 2-20   Low-frequency equivalent circuit for MOSFET.

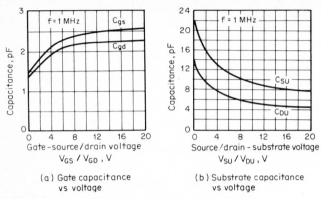

(a) Gate capacitance
vs voltage

(b) Substrate capacitance
vs voltage

FIGURE 2-21   MOSFET capacitance characteristics.

fringe effect becomes an important contribution to the feedback capacitance C_{gd}.

Because of the nonlinear nature of the FET, these equivalent-circuit elements are functions of the operating points; thus it is important that the bias conditions be specified when assigning values to the equivalent-circuit components.

2-8-1   Dual-Gate MOSFET

The dual-gate MOSFET is designed to decrease the degenerative feedback effect of C_{dg}. This structure in a typical application is equivalent to a common-source amplifier followed by a common-gate amplifier, an arrangement commonly called a "Cascode" circuit. Because of the low voltage gain of the input stage, the "Miller effect" of the C_{dg} is greatly reduced. The second gate also provides an effective means for automatic gain control (AGC). Application of the AGC signal to gate 2 results in a more remote cutoff characteristic than is achieved with a single-gate structure.

Figure 2-22a shows a section view of the dual-gate structure. This device is equivalent to two MOSFETs connected as shown in Fig. 2-22b. The more common symbol is given in Fig. 2-22c. A symbol with gate-protecting diodes is shown in Fig. 2-22d. The series diode arrangement shown permits both polarities of gate voltages—an important factor for some applications.

The output and forward transconductance characteristics of the dual-gate MOSFET are functions of both gates. If V_{GS2} is large, the output and transfer characteristics I_D vs. V_{GS1} and I_D vs. V_{DS} are similar to those of a single-gate device. Under these conditions, the voltage V_{GS2} has created a low-resistance channel between the drain and the internal

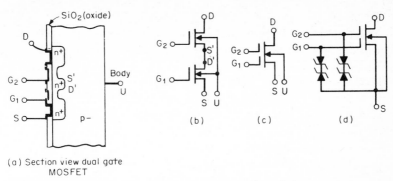

(a) Section view dual gate MOSFET

FIGURE 2-22   Dual-gate MOSFET—structure and symbols.

node $S'D'$. As V_{GS2} is decreased, the transconductance characteristic curves become saturated at a drain current determined by V_{GS2} as shown in Fig. 2-23a. The resulting forward transconductance characteristics are shown in Fig. 2-23b.

The small-signal equivalent circuit shown in Fig. 2-20 can also be utilized for the dual-gate MOSFET. The value of C_{dg} for the dual-gate unit is greatly reduced compared with an equivalent single-gate device, a typical value being 0.02 pF for the dual-gate unit compared with 1.0 pF for the single-gate unit.

2-8-2   Short-Channel MOSFET (VMOS)

The short channel and the backside drain contact of the VMOS (vertical-channel MOS) design result in higher current capability than has been achieved with conventional planar MOS designs. High drain-source voltage V_{DS} and low drain-gate capacitance C_{gd} characteristics are achieved by including a near-intrinsic n^- layer between the drain end of the chan-

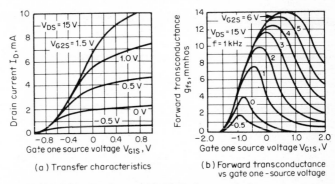

(a) Transfer characteristics

(b) Forward transconductance vs gate one-source voltage

FIGURE 2-23   Dual-gate MOSFET characteristics.

nel and the n^+ drain contact. The high V_{DS} advantage resulting from this design is achieved at the expense of drain-source symmetry; this means that drain and source functions cannot typically be interchanged. The source-body breakdown voltage is lower than drain-body breakdown. However, this nonsymmetry characteristic is unimportant in most applications where the body would be connected to the source. (For analog switch applications, Chapter 5 shows how VMOS units can be used back to back to achieve symmetrical high-voltage switching characteristics.) The short channel length causes a carrier velocity saturation effect which results in a low g_{os} as indicated in the output characteristics of Fig. 2-24a. The carrier velocity saturation also causes g_{fs} to saturate at ($V_{GS} - V_{GS(th)}$) values exceeding a few volts. For example, the transconductance g_{fs} of the VMOS type 2N6656 saturates at a V_{gs} value of about 4 V. This constant g_{fs} characteristic is indicated in the I_D-vs.-V_{gs} and the g_{fs}-vs.-V_{GS} curves of Fig. 2-24.

The transfer characteristic indicates that

$$I_D = K\,(V_{GS} - V_{GS(th)}) \tag{2-31}$$

where $V_{GS(th)}$ is the intercept at $I_D = 0$ of V_{GS} extrapolated from the linear portion of the characteristic (see Fig. 2-24b). Equation (2-31) also assumes that $V_{DS} > (V_{GS} - V_{GS(th)})$.

Gate capacitance of the VMOS may be separated into three components. The gate-to-drain region capacitance, at the bottom portion of the V groove, will decrease as V_{dg} is increased because of the increasing depletion of carriers in the n^- layer. Thus C_{rss} (or C_{gd}) has a negative V_{dg} coefficient and, because of the low carrier concentration in the n^- layer, will be fairly small.

Gate-to-source capacitance C_{gs} is a function of the gate-metal overlap into the source region. The carrier concentration in the n^+ source region is relatively high; therefore gate-source voltage has little effect upon the capacity of this region unless negative gate voltages are large enough to deplete the source. Positive gate-source voltages exceeding $V_{gs(th)}$ pull carriers into the channel region, increasing gate-to-channel capacitance. On either side of the threshold region this capacitance is not very voltage-dependent.

Drain-source capacitance C_{ds} is the junction capacitance of the drain-body diode. It thus has a negative voltage coefficient.

The gate capacitance of the typical VMOS structure, per unit of forward transconductance, is lower than that of the typical high-frequency JFET; thus excellent high-frequency performance can be achieved. Figure 2-24c shows capacitance characteristics of a VMOS device which has a g_{fs} of about 250 mmhos. The g_{fs}/C_{iss} is indicated as $(250 \times 10^{-3})/(50 \times 10^{-12}) = 5 \times 10^9$ mhos/F. For comparison the type U310

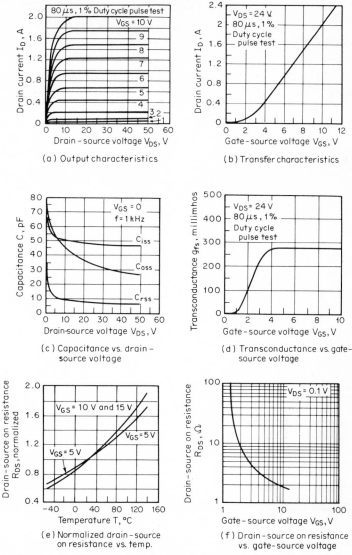

(a) Output characteristics

(b) Transfer characteristics

(c) Capacitance vs. drain–source voltage

(d) Transconductance vs. gate–source voltage

(e) Normalized drain–source on resistance vs. temp.

(f) Drain–source on resistance vs. gate–source voltage

FIGURE 2-24 **Performance characteristics—typical VMOS (2N6656).**

(a good 450-MHz JFET) has an indicated g_{fs}/C_{iss} ratio on the order of $10^{-2}/(5 \times 10^{-12}) = 2 \times 10^{9}$ mhos/F. A similar comparison using C_{rss} shows an even more significant advantage of VMOS.

VMOS FETs are available both with and without zener diodes or other protective devices shunting the gate-to-source voltage. The inclusion of these devices contributes to input admittance, which is undesirable for high-frequency or fast-switching applications. High-frequency, high-power VMOS FETs without gate-protecting diodes have been found to be relatively trouble-free if typical MOSFET handling precautions are exercised.

The positive temperature coefficient of the drain-to-source ON resistance (and the corresponding negative coefficient of I_D) serves as an aid in paralleling several devices for increased current and power capability. The "current-hogging" problem of paralleled bipolar transistors does not occur, and so negative-feedback resistors are not required.

2-9   ANALOG SWITCH PARAMETERS

The high ratio of OFF resistance to ON resistance of the FET has resulted in a variety of applications as electronic switches. The FET is especially useful in circuits which must control analog signals. In the ON state, a conducting channel connects source to drain with none of the junction barriers (and resulting offset voltages) that exist in diode, bipolar transistor, and SCR switches. In general, design goals for a FET switch are similar to those for a FET amplifier. That is, high channel conductance,

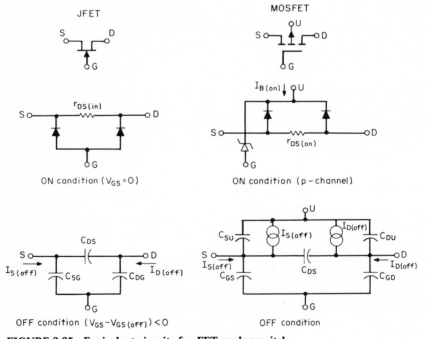

FIGURE 2-25   Equivalent circuits for FET analog switch.

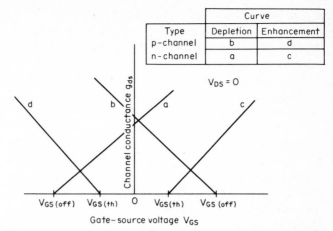

Type	Curve	
	Depletion	Enhancement
p-channel	b	d
n-channel	a	c

FIGURE 2-26 Channel conductance g_{ds} versus gate-source voltage V_{GS}.

low capacitance, low leakages, high breakdown voltage, and low gate-control voltages are desirable. For the circuit designer, however, different characteristic specifications are needed. This section covers the major switch parameters which are shown in the equivalent circuits of Fig. 2-25. The specifications assume that for the switch OFF condition the gate-source and gate-drain voltages are such that the channel is cut off. This means that for the n-channel FET the V_{GS} and V_{GD} must be more negative than the value of $V_{GS(off)}$ or $V_{GS(th)}$. For the switch ON condition, V_{GS} and/or V_{GD} must be more positive than the value of $V_{GS(off)}$ or $V_{GS(th)}$. In the ON condition the channel conduction is a function of gate voltage and drain-source voltage. Usually when the switch is ON, V_{DS} is low; therefore the ON resistance $r_{DS(on)}$ is specified at a low value of V_{DS}. Figure 2-26 shows channel conductance characteristics for n-channel and p-channel depletion-type and enhancement-type FETs.

Equations for these characteristics are as follows. For n-channel depletion type:

$$g_{ds} = \frac{1}{r_{ds}} = K\left(V_{GS} - V_{GS(off)}\right) \qquad (2\text{-}32a)$$

where $V_{GS} > V_{GS(off)}$

For p-channel depletion type:

$$g_{ds} = \frac{1}{r_{ds}} = K\left(V_{GS(off)} - V_{GS}\right) \qquad (2\text{-}32b)$$

where $V_{GS} < V_{GS(off)}$

For n-channel enhancement type:

$$g_{ds} = \frac{1}{r_{ds}} = K\,(V_{GS} - V_{GS(\text{th})}) \qquad\qquad (2\text{-}32c)$$

where $V_{GS} > V_{GS(\text{th})}$

For p-channel enhancement type:

$$g_{ds} = \frac{1}{r_{ds}} = K\,(V_{GS(\text{th})} - V_{GS}) \qquad\qquad (2\text{-}32d)$$

where $V_{GS} < V_{GS(\text{th})}$

[In Eqs. (2-32) the symbol $>$ means "more positive than" and $<$ means "less positive than."] As shown in Fig. 2-26 and Eqs. (2-32), making the n-channel V_{GS} more positive or the p-channel V_{GS} more negative (with respect to $V_{GS(\text{off})}$ or $V_{GS(\text{th})}$) will improve channel conduction and lower r_{ds}. For the JFET a "forward" bias at the gate-to-channel p-n junction should be avoided; therefore the ON resistance is normally specified with $V_{GS} = 0$.

For the enhancement-type MOSFET there is no gate-to-channel junction to limit V_{GS}. However, there is usually a gate-to-body protective diode included in the device. Also the gate-to-channel voltage should not exceed the maximum device ratings, or the gate dielectric may be damaged. In many applications, the source-to-body and drain-to-body voltage may be a function of the analog voltage being switched. Since the body acts like a "back" gate (with respect to the channel), the channel-to-body voltage will affect the value of $r_{ds(\text{on})}$. Figure 2-27 shows $r_{ds(\text{on})}$ characteristics for a Siliconix type M114, a p-channel enhancement-type MOSFET designed for analog switching applications. The specification for this device places limits on $r_{ds(\text{on})}$ at three different bias conditions as tabulated below.

Characteristic (M114)	Min	Max	Unit	Test conditions	
$r_{ds(\text{on})}$ drain-source		240	Ω	$V_{GS} = -40$ V, $V_{BS} = 0$	
ON resistance		275	Ω	$V_{GS} = -25$ V, $V_{BS} = 15$	$I_D = 0$
		500	Ω	$V_{GS} = -10$ V, $V_{BS} = 30$	$f = 1$ kHz

In the switch OFF state, the value of $(V_{GS} - V_{GS(\text{off})})$ or $(V_{GS} - V_{GS(\text{th})})$ must be negative for the n-channel FET or positive for the p-channel FET. Source and drain leakage is principally to the gate for the JFET and to the body for the MOSFET. These are reverse-bias p-n junction leakages and typically are proportional to the square root of the junction voltage.

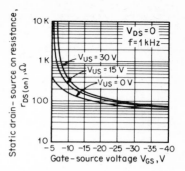

**FIGURE 2-27 Drain-source on re-
sistance vs. gate-source voltage.**

Source-to-drain feedthrough in the off state is a function of the capacitances shown in Fig. 2-25. C_{ds} is small (<0.1 pF) compared with
the others and can usually be neglected. Feedthrough via C_{SG}, C_{DG}, and
C_{SU} can be greatly reduced by providing a low-ac-impedance termination
for the gate and body terminals. Many data sheets quote input and
output capacitances for the common-source amplifier configuration, that
is, C_{iss}, C_{rss}, and C_{oss}. These parameters may be related to the circuits
of Fig. 2-25 by the equations

$$C_{iss} = C_{gs} + C_{gd} \qquad (2\text{-}33)$$

$$C_{oss} = C_{ds} + C_{gd} + C_{bd} \qquad (2\text{-}34)$$

$$C_{rss} = C_{gd} \qquad (2\text{-}35)$$

In addition to contributing to source-to-drain off-state feedthrough,
C_{gs} and C_{gd} will contribute to charge injection from the gate-control
circuit to the analog signal path.

Switching times (turn-on and turn-off times) are largely determined
by gate capacitance, which must be charged or discharged to change
the conduction state of the channel. A low value of capacitance is desirable for fast switching; however, if it is achieved by reducing channel
width, an increase in r_{ds} results. A good figure of merit is the $r_{ds} \times$
C_{iss} product, which should be small. The value of r_{ds} can be reduced
for a given geometry by increasing channel thickness; however, this also
increases $V_{GS(off)}$, and therefore, when comparing the $r_{ds} \times C_{iss}$ products
of various devices, the value of $V_{GS(off)}$ must also be considered. For
JFETs, $r_{ds(on)}$ is approximately proportional to $1/\sqrt{V_{GS(off)}}$ for a given
device geometry.

2-9-1 Offset Voltage

Ideally no currents are required to hold the FET switch in either the
on state or the off state. In the on state, there are no *p-n* junctions

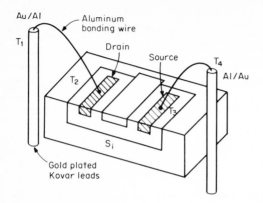

FIGURE 2-28 A typical system.

with their potential barriers in series with the conducting source-to-drain channel. However, offset voltages may occur from differential or unbalanced thermoelectric EMFs generated between source and drain. At each dissimilar metal junction, a thermoelectric EMF will exist. A typical system is shown in Fig. 2-28. In this system, the thermal offset voltage is given by

$$V_{(\text{thermal})} = (\text{Gold/Al coefficient}) \ (T_1 - T_4) + (\text{Al/}$$
$$\text{silicon coefficient}) \ (T_2 - T_3) \qquad (2\text{-}36)$$

The source and drain contacts on the FET chip are very close together, so that unless there is appreciable dissipation within the chip, $T_2 - T_3$ is very small. Thermal offset voltages of a discrete FET are typically less than 1 μV. If the device is a part of an integrated circuit or is included within a package containing heat-generating components which cause a thermal gradient across the chip, tens of microvolts of thermal EMF offset voltage may be developed.

2-10 TEMPERATURE EFFECTS

Changes in FET characteristics with increasing temperature can be related to one or a combination of the following: (1) a decrease in carrier mobility, (2) an increase in the thermal generation of hole-electron pairs, (3) a decrease in the depletion-layer width at p-n junctions, (4) a decrease in impact-ionization rates in high-field regions, and (5) an increase in donor and acceptor impurity ionization. These factors have the effect of causing most FET characteristics to improve with decreasing temperatures. Factor 5 is minor above about 70 K; however, below this, impurity ionization starts to decrease rapidly, so that JFET channel conduction also drops rapidly. The improvement in carrier mobility with decreasing temperature results in increasing channel conductance (and transconductance) until the 70-K region is approached. The increasing mobility also causes some reduction in avalanche breakdown of p-n junctions.

JFET gate "leakage" current I_G, in a normal common-source amplifier operating mode, is proportional to minority-carrier generation within the gate-junction depletion regions (space-charge regions). The two sources of carriers are those generated by thermal ionization and those generated by impact ionization. Thermal generation increases exponentially with temperature—doubling about every 10 or 12°C. Impact ionization decreases with temperature because it is a function of carrier velocity, which has a negative temperature coefficient. Figure 2-1d shows the typical net effect of these two characteristics upon I_G. In most applications where a low I_G is important, the FET will be biased below the "knee" or "breakpoint" of the I_G-versus-V_{DS} curve. If this is the case, then I_G will approximately double for every 10 to 12°C increase in temperature. Note that the breakpoint is a function of channel (drain) current. This is because the "excess" gate current is due to impact ionization by the carriers making up the channel current, and therefore is proportional to channel current.

The decrease in depletion-layer width at the p-n junctions results in an increase in channel thickness and thus an increase in $V_{GS(off)}$. This is equivalent to a V_{GS} change of about 2.2 mV/°C. (This is the same as the p-n diode forward-voltage change, and the base-emitter voltage change in a junction transistor.) With respect to channel conduction g_{ds} and drain saturation current I_{DSS}, the widening channel effect is opposed by a decrease in mobility. If $V_{GS(off)}$ is large (say 5 V), then the mobility decrease predominates and I_{DSS} and g_{DS} have a negative temperature coefficient of approximately $(1.006)^{-\Delta T}$. If $V_{GS(off)}$ is approximately -0.63 V, the channel thickness change exactly compensates the mobility decrease, which results in a near zero temperature coefficient for I_{DSS}. A FET with $V_{GS(off)}$ greater than 0.63 V can be biased down to its zero-TC operating point, at a V_{GS} of about $V_{GS(off)} + 0.63$ V. Figure 2-29 shows the I_D-versus-V_{GS} characteristics at three temperatures. It

FIGURE 2-29 FET transfer characteristics at three temperatures.

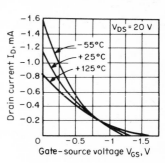

can be seen that for this device zero TC for I_D occurs at $V_{GS} \cong 0.9$ V, which is about 0.63 V above $V_{GS(off)}$.

Temperature effect upon gate capacitance is a function of how the device is biased. The change in depletion-layer thickness tends to give the junction capacitance a positive coefficient. However, if the bias is large (drain-gate voltage, for example), then the percentage change in depletion-layer thickness (and thus the capacitance) is small.

2-11 DERATING FACTORS

As shown above, the FET electrical characteristics are a function of temperature. The temperature of importance is that of the active region of the device, which probably will be higher than the surrounding ambient temperature due to internal heating. Every part of the FET has thermal properties: the silicon die, the die attach material (solder, epoxy, etc.), and the header. The thermal properties include thermal resistance and capacitance, analogous to electrical resistance and capacitance. The die temperature rise above ambient is the product of the die power dissipation and the thermal resistance between the die and the ambient. Thermal resistance is normally expressed as °C/W (degrees Celsius per watt). For power devices, the temperature rise is usually specified with reference to the device case (package) temperature; thus it does not take into account the thermal resistance involved in mounting the case to a heat sink.

Temperature is important from the standpoints of electrical characteristics and long-term reliability. The maximum device temperature and dissipation ratings specified in data sheets usually are related to long-term reliability considerations. Recommended maximum internal temperatures established by users and manufacturers of silicon FETs are usually in the 150 to 200°C range. Dissipation ratings are thus related to the recommended maximum device temperature. For example, assume a device with a maximum temperature rating of 200°C has a thermal resistance of 30°C/W. If the case temperature is held to a maximum of 50°C, then the dissipation should be limited to

$$P = \frac{T_2 - T_1}{°C/W} = \frac{200 - 50}{30/W} = 5 \text{ W}$$

For another example, assume a device has a "free-air" (no heat sink) power dissipation derating factor of 3 mW/°C. This implies that the thermal resistance is $(3 \text{ mW}/°C)^{-1}$ or 333°C/W. If the device is operated at a power dissipation level of 50 mW, then the internal temperature T_j will be

$$\Delta T_j = (0.050 \text{ W}) \frac{(333°C)}{\text{W}} = 16.5°C$$

above the ambient temperature. The free-air rating means that the thermal resistance of case to surrounding air is included in the rating, that is, no heat sink is assumed.

2-12 GLOSSARY OF TERMS AND ABBREVIATIONS

BV_{GSS}	Gate-source breakdown voltage
C_{bd}	Body-drain capacitance
C_{gd}	Gate-drain feedback capacitance
C_{gs}	Gate-source capacitance
C_{iss}	Common-source input capacitance
C_{oss}	Common-source output capacitance
C_{rss}	Common-source reverse transfer capacitance
C_{ds}	Drain-source capacitance
$\bar{e}_n$	Equivalent short-circuit input noise voltage
$\bar{i}_n$	Equivalent short-circuit input noise current
g_{fs}	Common-source forward transconductance
g_{fso}	Common-source forward transconductance when $V_{GS} = 0$
g_{iss}	Common-source input conductance
g_{os}	Common-source output conductance
$I_{D(off)}$	Drain cutoff current
I_D	Drain current
I_{DSS}	Saturation drain current (at $V_{GS} = 0$)
I_G	Operating gate current
I_{GSS}	Gate reverse current
$r_{ds(on)}$	Dynamic drain-source ON resistance
$r_{DS(on)}$	Static drain-source ON resistance
R_L	Drain load impedance
V_{DS}	Drain-source voltage
V_{DG}	Drain-gate voltage
V_{GS}	Gate-source voltage
V_p	Pinchoff voltage
$V_{GS(th)}$	Gate threshold voltage
$V_{GS(off)}$	Gate-source cutoff voltage

REFERENCES

1. Dacey, G. C., and I. M. Ross, "Unipolar Field-Effect Transistor," *Proceedings, IRE,* **41**:970–979, 1953.

2. Middlebrook, R. D., "A Simple Derivation of Field-Effect Transistor Characteristics," *Proceedings, IEEE,* **50:**1146–1147, 1963.

3. Nyquist, H., "Thermal Agitation of Electric Charge in Conductors," *Phys. Review,* **32:**110, 1928.

4. Radeka, V., "Field-Effect Transistors for Charge Amplifiers, *IEEE Trans. Nucl. Sci.,* **NS-20**(1), 182–189, 1973.

5. Shockley, W., "A Unipolar Field-Effect Transistor," *Proceedings, IRE,* **40:**1365–1376, 1952.

3

LOW-FREQUENCY CIRCUITS

3-1 Introduction
3-2 Gross Boundaries of the Operating Region
3-3 Design Example
3-4 Establishing the Operating Point and Bias Method
3-5 Input Capacitance
3-6 Choosing the FET—A Design Example
3-7 Constant Current-Source Bias
3-8 Cascode Circuits
3-9 Source Follower (Common-Drain Amplifier)
3-10 Differential Amplifier
3-11 Distortion in FET Amplifier
3-12 Audio-Frequency Noise Characteristics

3-1 INTRODUCTION

FET electrical characteristics were discussed in Chapter 2, and equivalent circuit models were developed. In this chapter we will utilize these characteristics and models to develop FET amplifier configurations.

Three basic problems facing the designer of an amplifier are (1) the choice of the circuit configuration, (2) the selection of the active device type to be utilized, and (3) the setting of the bias conditions. The choices for an amplifier stage that delivers 50 W of audio power to a speaker differ from those for one which amplifies a 5-μV signal from a high-impedance source. The selection of a device and its optimum bias are interrelated; however, the approximate bias can be determined prior to the selection of a specific device type.

3-2   GROSS BOUNDARIES OF THE OPERATING REGION

For a linear amplifier there are gross boundaries on the FET output characteristics within which the operating point (bias) must be located. For reliable operation the voltage, current, and power-dissipation ratings of the FET should not be exceeded. For low distortion V_{DG} much below the I_D saturation region should be avoided. The minimum drain-voltage boundary should be such that[*]

$$V_{DG} = -V_p \qquad (3\text{-}1)$$

The maximum drain voltage should be specified so that the drain-gate or drain-source breakdown voltage is not likely to be exceeded. The maximum drain current may be limited by the I_D (max) rating of the device or typically, in the case of a small-signal device, by the I_{DSS} value. I_D may be permitted to be greater than I_{DSS} if the application permits a forward bias on the gate.

The maximum power-dissipation rating results in the boundary of the maximum $V_{DS} \cdot I_D$ product. This boundary is a function of maximum operating temperature. In some cases exceeding the dissipation boundary may be permitted if the time factor is such that the rise in device temperature does not exceed the device maximum temperature rating.

The shaded area of Fig. 3-1 shows the allowed operating region. The quiescent point (zero-signal bias point) must be such that with

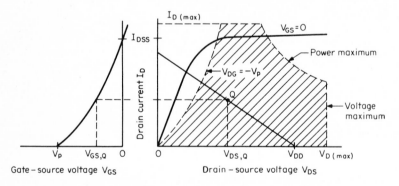

FIGURE 3-1   Gross operating boundaries for linear operation.

the maximum required output voltage and current swing, V_{DG} is not forced below $-V_p$ and I_D is not forced to zero. Further restrictions on the gross boundaries may be set by other factors such as available power-supply voltage and power consumption limits.

---

[*] We have chosen to let $V_p = V_{GS(off)}$. This differs from some writers' use of V_p to mean the value of V_{DS} at which I_D saturates. For the classical FET, with characteristics as shown in Fig. 2-1, I_D saturation occurs when $V_{DG} = -V_p$.

3-3 DESIGN EXAMPLE

We will design a FET voltage-amplifier stage and discuss some steps in choosing a suitable FET. Using this FET amplifier, we will then discuss biasing methods. For the voltage-amplifier stage shown in Fig. 3-2, make the following assumptions:

1. Voltage needed to drive the VMOS power output stage
$$e_o = 2 \text{ V rms} = 2.8 \text{ V peak}$$

2. High-frequency (-3 dB) corner ≥ 50 kHz

3. $R_L = 10^6 \ \Omega$

4. $C_L = 200$ pF

5. Total available supply voltage $= 30$ V

6. $R_{\text{gen}} = 30{,}000 \ \Omega$

7. Low-frequency corner ≤ 20 Hz

8. Voltage gain e_o/e_{gen} to be as high as practical

There are three amplifier configurations: the common-drain, the common-gate, and the common-source. The common-drain stage has a very high input impedance and a low output impedance; however, its voltage gain is less than 1. The common-gate stage can have a high voltage gain, but its input impedance is low. The common-source stage has both high input impedance and high voltage gain, so it is the configuration we will use for this amplifier example. We will start with the circuit of Fig. 3-2. The approximate equivalent circuits for low frequency, mid-frequency, and high frequency are shown in Fig. 3-3.

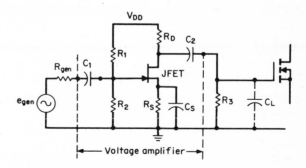

FIGURE 3-2 JFET voltage amplifier stage.

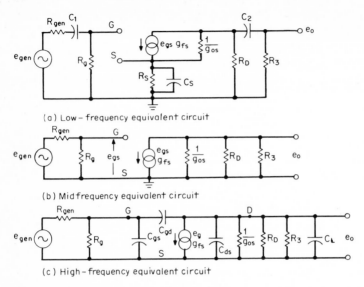

(a) Low–frequency equivalent circuit

(b) Midfrequency equivalent circuit

(c) High–frequency equivalent circuit

FIGURE 3-3   Equivalent circuit of Fig. 3-2.

3-4   ESTABLISHING THE OPERATING POINT AND BIAS METHOD

The allowed operating region has been indicated in Fig. 3-1. How do we establish the quiescent operating point and how accurately need it be maintained? Fortunately, biasing of a single-stage FET is a relatively simple matter as compared with a bipolar transistor. The input (gate) of the FET draws very little direct current, and FET dc characteristics are not as temperature-dependent as bipolar characteristics. However, parameters do vary with temperature. Also, variations will occur from device to device within the same type number. To obtain fairly uniform performance over a wide temperature range with devices of a given type, it is desirable to control I_D. When biased to a given I_D, g_{fs} varies less from unit to unit and with temperature than would be the case if it were biased with a constant V_{GS}. If the maximum possible output voltage is needed, then constant I_D biasing is the best method to use. Constant V_{GS} biasing is seldom used in low-frequency RC-coupled amplifier circuits because of the wide spread in I_{DSS} usually specified in device types.

Biasing in such a way that I_D is a controlled function of V_p may be even better than the constant-I_D method. With constant current biasing, units with a high V_p typically will have a lower g_{fs} than will units of the same device type with a low V_p. By permitting the I_D to increase for the high-V_p units, the g_{fs} can be held more nearly constant. This will result in a change in V_{DG}, which must be taken into account when considering the maximum output.

Figure 3-4 shows one method of making I_D a function of V_p. The source current is supplied through a source resistor R_S, and V_G is constant. This is sometimes called the self-bias method and is widely used for ac amplifiers.

For the forward-transfer and the output characteristic curves shown in Fig. 3-1, the source terminal was used as a reference (V_s was constant). In the circuit of Fig. 3-2, however, the gate voltage V_G is constant (assuming I_G is negligible). To study this bias situation, a different presentation of the characteristics will be used. For our design we will maintain $V_{DG} > -V_p$ so that to a first approximation I_D is affected only by V_{SG}. Figure 3-4 shows the I_D-versus-V_{SG} characteristic curves for two devices that we will consider to be our limiting units. Q_1 is the unit with maximum V_p, and Q_2 is the unit with minimum V_p. Our goal is to minimize g_{fs} variations between the limit devices. We will try to bias Q_1 so that its g_{fs} at $I_{D(1)}$ is the same as the Q_2 g_{fs} at $I_{D(2)}$. The maximum value of $I_{D(1)}$ to avoid output clipping is

$$I_{D(1)} = \frac{V_{DD} - V_G + V_P - \sqrt{2}\, e_o}{R_D} \qquad (3\text{-}2)$$

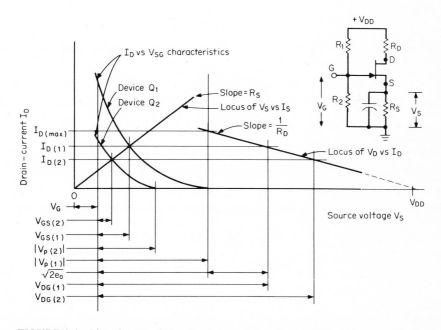

FIGURE 3-4 **Bias characteristics for the amplifier circuit of Fig. 3-2.**

The problem is a little more complicated now because of our desire to maintain a constant g_{fs}. It can be seen in Fig. 3-4 that for a given $I_{D(1)}$, V_G and $I_{D(2)}$ will be a function of R_S. We want $I_{D(2)}$ to be the value that results in the g_{fs} of Q_2 being the same as the g_{fs} of Q_1.

To determine $I_{D(2)}$ we make use of Eqs. (2-12) and (2-10). By letting $g_{fs(1)} = g_{fs(2)}$, we can get $I_{D(2)}$:

$$I_{D(2)} = I_{D(1)} \frac{I_{DSS(1)}}{I_{DSS(2)}} \left(\frac{V_{p(2)}}{V_{p(1)}} \right)^2 \qquad (3\text{-}3)$$

where the subscripts (1) and (2) indicate devices 1 and 2.

To obtain the slope R_S, we need the values of $V_{SG(1)}$ and $V_{SG(2)}$, which may be derived from Eq. (2-1).

$$V_{GS} = V_p \left[1 - \left(\frac{I_D}{I_{DSS}} \right)^{1/2} \right] \qquad (3\text{-}4)$$

Then from Fig. 3-4 we get

$$R_S = \frac{\Delta V_S}{\Delta I_S} = \frac{V_{SG(1)} - V_{SG(2)}}{I_{D(1)} - I_{D(2)}} \qquad (3\text{-}5)$$

and

$$V_G = R_S I_{D(2)} - V_{SG(2)} \qquad (3\text{-}6)$$

We can estimate the g_{fs} at the operating I_D with an equation derived from Eqs. (2-10) and (2-12):

$$g_{fs} = \frac{2(I_D I_{DSS})^{1/2}}{-V_p} \qquad (3\text{-}7)$$

The value of R_D is needed to determine $I_{D(1)}$ and the voltage gain A_V. To maximize A_V, R_D should be high; however, an upper limit is imposed by the required frequency response. The output side of the equivalent circuit shown in Fig. 3-3 will have a high-frequency corner which will be a function of R_o. Midfrequency output voltage is

$$e_{o(MF)} = -e_g g_{fs} R_o' \qquad (3\text{-}8)$$

High-frequency output is

$$e_{o(HF)} = \frac{-e_g g_{fs} R_o'}{[1 + (2\pi f C_L R_o')^2]^{1/2}} \qquad (3\text{-}9)$$

where $R_o' = \dfrac{1}{1/R_D + 1/R_G + g_{os}}$ $(3\text{-}10)$

The fall-off in output at high frequency resulting from C_L is obtained by the ratio of Eq. (3-9) to Eq. (3-8):

$$\frac{e_{o(\text{HF})}}{e_{o(\text{MF})}} = \frac{1}{[1 + (2\pi f C_L R_o')^2]^{1/2}} \tag{3-11}$$

(assuming e_g is constant).

3-5 INPUT CAPACITANCE

On the input side of the high-frequency equivalent circuit, the capacitance C_{in} will cause e_g to decrease as frequency is increased. The equation to determine this loss is similiar to Eq. (3-11):

$$\frac{e_{g(\text{HF})}}{e_{g(\text{MF})}} = \frac{1}{[1 + (2\pi f C_{\text{in}} R_G')^2]^{1/2}} \tag{3-12}$$

where R_G' is the parallel combination of R_{gen}, R_1, and R_2, and

$$C_{\text{in}} = C_{gs} + C_{gd}(1 - A_V) \tag{3-13}$$

C_{gs} and C_{gd} are functions of V_{GS} and V_{GD}. Therefore we must determine bias conditions as well as A_V to determine C_{in}.

Junction capacitance C_j as a function of voltage has the form

$$C_j = C_o \left[1 + \left(\frac{V_j}{V_{bi}}\right)^k\right]^{-1} \tag{3-14}$$

If C_j is known at one bias voltage $V_{j(1)}$, it can be determined at another bias voltage $V_{j(2)}$ by

$$C_{j(2)} = C_{j(1)} \left(\frac{V_{bi} - V_{j(1)}}{V_{bi} - V_{j(2)}}\right)^k \tag{3-15}$$

For an abrupt junction, $k = 0.5$; for a linearly graded junction, $k = 0.333$. Most JFET gate junctions fall between these two conditions, so we will use $k = 0.4$. For silicon the "built-in" space-charge voltage V_{bi} is about 0.6 V. The FET specification sheet will normally give the bias condition for the C_{gs} and C_{gd} specs. For the amplifier, V_{GS} is given by Eq. (3-4). By observation from Fig. 3-4 we can get

$$-V_{GD} = V_{DD} - V_G - I_D R_D \tag{3-16}$$

3-6 CHOOSING THE FET—A DESIGN EXAMPLE

We will use these equations in the design of the amplifier of Fig. 3-2. A point that should be emphasized is that not all JFET device types conform to the equations presented in Chapter 2 and in this chapter.

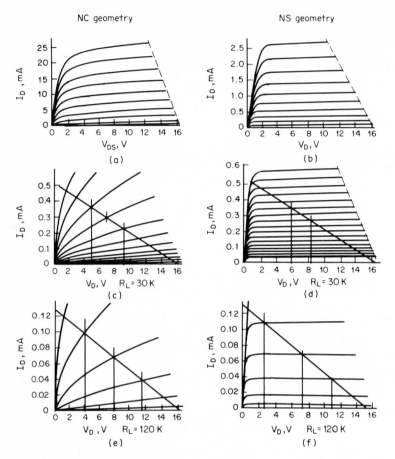

FIGURE 3-5 Comparison of short-channel FET (NC geometry) with long-channel FET.

After a part number has been selected for the amplifier, its characteristics should be checked with the design operating conditions, for conformance with the design goals. For example, the output characteristics of two different FET geometries are given in Fig. 3-5. The active chip areas of these two geometries are approximately equal, and their $V_{GS(off)}$ values are about the same. It can be seen, however, that the NC geometry has an I_{DSS} about 10 times that of the NS geometry. This is due to the difference in the channel length and width of the two types, the shorter length of the NC channel resulting in a higher I_{DSS} and g_{fs} per unit of channel width. It also results in a higher g_{fso}/C_{gd} ratio, which is an important factor for high-frequency amplifiers and for low-ON-resistance analog switches. The long-channel NS geometry, on the other hand, is good for low-noise performance at low frequencies where its relatively low g_{fs}/C_{gd} ratio is not so important as it is for high-frequency devices. Figure 3-5 compares the output characteristics of these two devices at various I_D levels.

The short-channel effect of the NC geometry is very noticeable at low-current-bias conditions. The high g_{os} is apparent. It can also be seen, however, that when biased for the same I_D condition, the g_{fs} is much higher for the NC than for the NS. Even with the higher g_{os}, the A_V [Eq. (2-13)] for the NC unit can be greater than for the NS unit. These characteristics should point out the importance of not taking for granted all equation simplifications. Neglecting g_{os} [Eq. (2-14)] to arrive at a simplified equation for A_V would cause little error for the NS geometry with the load lines shown in Fig. 3-5; however, it would result in a large error with the NC geometry.

Deciding what FET type to use for a given application can be a problem. There are several hundred part numbers listed in a typical FET catalog. Fortunately almost all these numbered types are made from just a few basic production designs. These designs are referred to as "geometries." Within a given geometry the I_{DSS} and $V_{GS(off)}$ are related approximately as indicated by Eqs. (2-1) through (2-4). The major difference between devices within a given geometry is the channel thickness, usually controlled by an epitaxial-layer thickness, ion implant depth, diffusion time, or a combination of these. Figure 3-6 shows the I_{DSS}

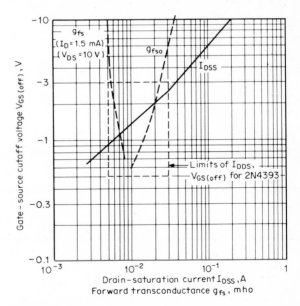

FIGURE 3-6 I_{DSS} **and** g_{fs} **versus** $V_{GS(off)}$ **relationship of NC geometry.**

versus $V_{GS(off)}$ relationship of devices made from Siliconix Inc.'s "NC" geometry. About 40 "standard" part numbers are characterized from this one geometry.

For our amplifier example we will utilize a "JFET geometry selector guide" as was shown in Fig. 2-17. Certain parts of this guide are reproduced in Fig. 3-7 for convenience. The curves indicate a V_p range from about 0.2 to 10 V. With our knowledge that higher A_V can be achieved with lower V_o units (provided I_D is adequate), we tend to favor these units for our design. Since 3 V is about the middle of the V_p range indicated, we will use this as a tentative value.

Figure 3-4 shows the bias characteristics for the amplifier. The bias current $I_{D(1)}$ can be calculated with Eq. (3-2); however, we must first set values for R_D and V_G. Our amplifier design called for e_o to be 2 V rms, and we have assumed $V_p = -3$ V. The upper limit for R_D is determined by the maximum permissible fall-off at high frequency as indicated by Eq. (3-11). The design goal called for less than 3 dB loss at 50

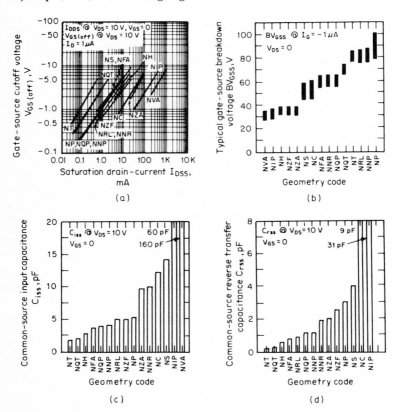

FIGURE 3-7   JFET geometry selector guide.

kHz. There will be some loss at the input due to C_{in}; however, we can't determine C_{in} until we know the device we are going to use and the value of R_D. We will assume for the moment a 1.5-dB loss at the input and a 1.5-dB loss at the output.

A loss of 1.5 dB is a voltage ratio of 0.841. Using this value we use Eq. (3-11) to obtain a value for R_o:

$$R_o = \frac{[(e_{o(MF)}/e_{o(FH)})^2 - 1]^{1/2}}{2\pi f C_L} \qquad (3\text{-}17)$$

With $f = 50$ kHz, $C_L = 200$ pF, and the 1.5-dB loss indicated above, we get with Eq. (3-17) a value of $R_o = 10,239$ Ω. R_o is the parallel combination of R_D, R_L, and $1/g_{os}$. In our example $R_L = 1$ MΩ, so it can be neglected. Figure 3-5 shows that g_{os} is a function of the FET geometry, so we cannot get its value until we select the device type. For the moment we will assume that $1/g_{os} \gg R_L$ and let $R_L = R_o$. We will modify this after we determine the approximate g_{os} for the selected device type.

The best value for V_G will be a function of the geometry; for now we will assume a value of 3 V so that we can proceed with estimating an $I_{D(1)}$ value. Using Eq. (3-2) we get

$$I_{D(1)} = \frac{30 - 3 - 3 - 2.8}{10,239} = 2 \text{ mA}$$

Minimum I_{DSS} of the selected FET should allow for a peak signal current above the $I_{D(1)}$ value.

$$I_{D(max)} = I_{D(1)} + \frac{\sqrt{2}\, e_o}{R_D} \qquad (3\text{-}18)$$

If $I_{DSS(min)}$ is lower than $I_{D(max)}$, then the gate will become forward-biased on positive input peaks. Typically I_{DSS} will decrease as temperature increases, so this must be taken into account. The maximum temperature coefficient for I_{DSS} is approximately $e^{-(0.006)\Delta T}$. If we assume a maximum operating temperature of 75°C, then the 25°C value of I_{DSS} should be

$$I_{DSS(25°)} = I_{DSS(75°)} e^{-(0.006)(25-75)} = 1.35 I_{DSS(75°)}$$

With $I_{D(1)} = 2$ mA, $I_{D(max)}$ per Eq. (3-18) will be

$$I_{D(max)} = 2 + \frac{(1.414)(2)}{10.239} = 2.28 \text{ mA}$$

and $I_{DSS(25°)} = (1.351)(2.28) = 3.08$ mA (min).

Looking at the selector guide in Fig. 3-7 we tabulate in Table 3-1 FET geometries having $I_{DSS} > 3$ mA. We have added to this table the approximate V_p, BV_{GSS}, and capacitance values.

TABLE 3-1   SELECTED FET GEOMETRIES HAVING $I_{DSS} > 3$ mA

Geometry	Approximate $-V_P$ for $I_{DSS} = 3$ mA	BV_{GSS} range	At $V_{DS} = 10$ V and $V_{GS} = 0$	
			$C_{iss} - C_{rss}$	C_{rss}
NS	2.7 V	>40 V	10 pF	4 pF
NFA	3.3	>40	3.5	0.8
NP	2.3	>40	3.5	1.6
NRL	1.5	>40	4	1
NH	1.3	30–40	2	0.6
NZF	0.9	30–40	5	1
NC	<0.5	>40	10	3
NZA	<<0.5	25–35	8	2
NVA	<0.5	25–35	160	—

NOTE: NQT, NQP, NNP, and NNR geometries are dual FETs and so they are not considered for this application.

High gain at midfrequency could be achieved with the NVA geometry; however, because of its large capacitance values, the high gain would probably have to be sacrificed to get the desired high-frequency corner. Also there are no standard devices from NVA geometry having a V_{DG} rating of 30 V or greater, so we will not evaluate this geometry further.

The NC geometry is next in line with respect to I_{DSS}/V_p ratio. It has a high BV_{GSS} and reasonably low C_{iss} and C_{rss}. There are several devices using this geometry that have low V_p specs of -0.5 to -3 V. We will select the 2N4393 for further evaluation.

The NZF geometry has lower C_{iss} and C_{rss} than the NC; however, there are no standard device types with BV_{GSS} of 30 V or greater.

The low-capacity NH geometry should be further evaluated and compared with the device selected from the NC geometry. We find that the FET type 2N4416 has a 5-mA I_{DSS} minimum, and also has a 30-V BV_{GSS} minimum, so we select it for evaluation. The other geometries in Table 3-1 have higher V_p and higher C_{iss} than the NH, so they will not perform as well.

Table 3-2 is used to list the characteristics of the 2N4416 and the calculated amplifier-component values. In the "Note" column we indicate whether the listed characteristic is a data-sheet limit, a value estimated from the data-book curves, or a calculated value. We have a column for the high-V_p unit (Q_1 in Fig. 3-4) and a column for the low-V_p unit (Q_2 in Fig. 3-4). The performance curves will aid us in estimating

TABLE 3-2 DEVICE TYPE 2N4416 (SILICONIX NH GEOMETRY)

Characteristic	Value for Q_1 (high V_p)	Value for Q_2 (low V_p)	Unit	Note
I_{DSS}	15	5	mA	1
V_p	6	2.5	V	1
C_{gs}	3.2	3.2	pF	$V_{os} = 0$
C_{gd}	0.8	0.8	pF	$V_{gd} = -15$ V
I_{DSS}	15		mA	1
I_{DSS}		6	mA	2
V_p	4.8		V	2
V_p		2.5	V	1
g_{os}	0.013	0.007	mmho	2
R_D	12	12	kΩ	Eq. (3-17)
R_o	10.4	11	kΩ	Eq. (3-10)
I_D	1.5	1.02	mA	Eqs. (3-2), (3-3)
V_{SG}	3.28	1.47	V	Eq. (3-4)
R_S	3.77	3.77	kΩ	Eq. (3-5)
V_G	2.38	2.38	V	Eq. (3-6)
g_{fs}	3	3	mmho	2
e_g / e_g	31.2	33		$g_{fs} R_o$
V_{DG}	9.62	15.38	V	Eq. (3-16)
C_{gs}	1.55	1.95	pF	Eq. (3-15)
C_{gd}	0.95	0.8	pF	Eq. (3-15)
$e_{o(HF)} / e_{o(MF)}$	0.837	0.823	—	Eq. (3-11)
$e_{g(HF)} / e_{g(MF)}$	0.957	0.965	—	Eq. (3-11)
$e_{o(HF)} / e_{g(MF)}$	0.801	0.794	—	
$20 \log [e_{o(HF)} / e_{g(MF)}]$	−1.93	−2.0	dB	
C_1	0.03		μF	Eq. (3-20)
C_2	0.03		μF	Eq. (3-20)
C_3	33		μF	Eq. (3-20)
R_1	12		mΩ	Eq. (3-21)
R_2	1		mΩ	Eq. (3-21)

NOTES:
1. Data-sheet limit.
2. Estimated from curves.
3. Calculated with equation.

the interrelation of the various characteristics. For example, we note that the typical $I_{DSS(\text{max})}$ unit of the 2N4416 (NH geometry) has a V_p of about -4.8 V instead of the -6-V limit permitted by the specification sheet.

Also the typical $V_{p(\text{min})}$ unit has an I_{DSS} of about 6 mA instead of the data-sheet limit of 5 mA.

In calculating $I_{D(1)}$ we will use -6 V for V_p because this will ensure that a data-sheet-limit unit will not result in output clipping at maximum e_o. For calculating $I_{D(2)}$, however, we will use the relationships indicated by the performance curve. This should result in better selection of the proper R_S for g_{fs} control.

We proceed with the calculations and fill in Table 3-2. In arriving at the first approximation for $I_{D(1)}$, g_{os} was neglected. Now that we have I_D, we can estimate from the performance curves g_{os} for the two devices. For the NH with high V_p we get $g_{os} = 0.013$ mmho.

The calculated value of g_{fs} is 1.98 mmhos. The g_{fs} versus I_D performance curve in the data book indicates that g_{fs} is closer to 3 mmhos for this device. The error in the equation is probably due to the short-channel effect in the NH geometry. We used the higher of the two values in Table 3-2.

The midfrequency gain (e_o/e_g) is slightly higher for the low-V_p unit because the lower I_D gives a lower g_{os}. This could be compensated by lowering $I_{D(2)}$ to reduce $g_{fs(2)}$.

The high-frequency loss is only 2 dB because the C_{in} loss is less than we initially allowed for. Figure 3-3a shows the low-frequency equivalent circuit. Response will be determined by the input and output coupling capacitors, and by the source-bypass capacitor. The low-frequency transfer function of a simple RC high-pass filter is

$$K_{LF} = \frac{1}{\{1 + [1/(2\pi fCR)]^2\}^{1/2}} \tag{3-19}$$

We have three RC networks to consider: (1) the input, consisting of C_1, R_{gen}, and R_g'; (2) the output, consisting of C_2, R_D, g_{os}, and R_3; and (3) the source circuit, consisting of C_S, R_S, and g_{fs}.

We let the attenuation resulting from these three RC networks be -3 dB, i.e., the product of the three K_{LF} factors be 0.708. Since the input and output R_S are much larger than the equivalent source resistance, C_1 and C_2 can be much smaller than C_S. If we let K_{LF} associated with C_1 and C_2 each $= 0.95$, then K_{LF} associated with C_S will be

$$K_S = \frac{0.708}{(0.95)(0.95)} = 0.784$$

where K_S is low-frequency attenuation resulting from imperfect source bypassing. For both the input and the output, R is approximately 10^6 Ω. Assuming $K = 0.95$ at $f = 20$ Hz,

$$C = [2\pi f R (K_{LF}^{-2} - 1)^{1/2}]^{-1} \tag{3-20}$$

Using the above values we get C_1 and $C_2 = 0.0242$ μF. We will use 0.03 μF.

For the source circuit, the equivalent R is R_S in parallel with g_{fs}^{-1}. Using Eq. (3-20) and letting $K = 0.784$, we get $C_S = 33$ μF. The voltage across C_S will not exceed $R_S/(R_D + R_S)$ (V_{DD}). A 10-V rating for C_S is used.

We have assumed the parallel combination of R_{G1} and $R_{G2} = 10^6$ Ω. The ratio is set by V_{DD} and V_G.

$$\frac{R_{G2}}{R_{G1} + R_{G2}} = \frac{V_G}{V_{DD}} \tag{3-21}$$

Since we want

$$\frac{R_{G2}}{R_{G1} + R_{G2}} = 10^6 \ \Omega$$

and

$$\frac{R_{G2}}{R_{G1} + R_{G2}} = \frac{-V_G}{V_{DD}} = \frac{2.38}{30}$$

$$R_{G(1)} = (10^6)\left(\frac{30}{2.38}\right) = 12.6 \ \text{M}\Omega$$

Use 12 MΩ. Then

$$R_{G2} = \frac{2.38}{30} (R_{G1} + R_{G2})$$

$$R_{G2} = \frac{2.38}{30} R_{G1}\left(1 - \frac{2.38}{30}\right)^{-1} = 1.03 \ \text{M}\Omega$$

Use 1 MΩ.

The completed amplifier design is shown in Fig. 3-8.

3-7 CONSTANT CURRENT-SOURCE BIAS

Replacing R_S in Fig. 3-8 with a current source results in the circuit and bias conditions shown in Fig. 3-9. The value of the current source

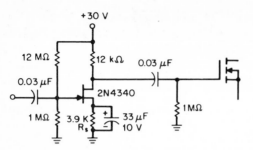

FIGURE 3-8 Completed amplifier schematic.

I_S determines the value of I_D. If I_S is constant, then $I_{D(1)} = I_{D(2)}$. I_D does not change from device to device. From a practical standpoint a FET current-limited diode, such as the Siliconix type CR100, can be used. By setting V_G to $+2$ V, the CR100 operates well into its saturated region, so that its effective output resistance is greater than 180 kΩ. Hence, a V_{SG} change of 2 V results in a less than 12-μA change in I_D. However, we must take into account the current-source tolerance. If we assume a ± 10 percent tolerance, for example, then the I_{DQ} range would be 0.9 to 1.1 mA. As with the source resistance example, the current source must be bypassed with C_S of essentially the same value.

An advantage of the constant current source over the large source resistor is the improvement in current stability. The drain current is essentially independent of the I_{DSS} and V_p of the FET (within limits). As shown in Fig. 3.9, $I_{D(1)}$ and $I_{D(2)}$ are equal to the current of the regulator CR100. With CR100 limits of 0.9 and 1.1 mA, V_D would be

$$V_{D(\text{high})} = 30 \text{ V} - (0.9 \text{ mA})(12 \text{ k}\Omega) = 19.2 \text{ V}$$
$$V_{D(\text{low})} = 30 \text{ V} - (1.1 \text{ mA})(12 \text{ k}\Omega) = 16.8 \text{ V}$$

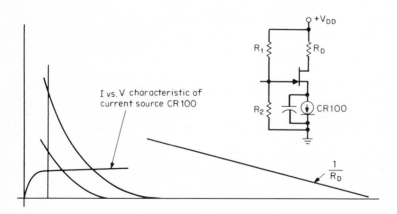

FIGURE 3-9 Characteristics with constant current bias.

Maximum output voltage without clipping would be

$$e_{o(\text{high}) \text{ max}} = 30 - 19.2 = 10.8 \text{ V}$$
$$19.2 - 6 = 13.6 \text{ V}$$
$$e_{o(\text{low}) \text{ max}} = 30 - 16.8 = 13.2 \text{ V}$$
$$16.8 - 6 = 10.2 \text{ V}$$

For the worst case,

$$e_{o(\text{peak})} = 10.2 \text{ V}$$

An important application of the current regulator (or "current source") is in differential amplifiers. The high output resistance of the current source enhances the common-mode rejection.

3-8 CASCODE CIRCUITS

The term "Cascode" describes an amplifier stage consisting of a common-source (or emitter or cathode) stage, followed by a common-gate (or base or grid) stage. This circuit configuration was originally developed to achieve improved high-frequency performance in vacuum tube amplifiers. The Cascode configuration not only improved the output-to-input isolation; it also greatly improved the high-frequency amplifier stability by reducing the feedback capacitance within the amplifier stage. In this section Cascode circuit advantages in dc and low-frequency amplifier circuits are covered. High-frequency circuits are discussed in Chapter 4.

3-8-1 Increasing Junction-FET-Amplifier Input Resistance by Using Cascode Circuits

The gate-leakage current I_G as a function of drain-to-gate voltage V_{DG} discussed in Chapter 2 increases rapidly once the I_G breakpoint voltage is exceeded (Fig. 2.1d). Cascode circuits are useful in amplifier applications requiring high input resistance over a wide input-voltage range; for example, source followers and differential amplifiers. Before discussing the circuits, let's look a little closer at the I_G breakpoint mechanism.

3-8-2 I_G Breakpoint Mechanism

The gate current I_G of an n-channel JFET is a function of gate-source voltage V_{GS}, gate-drain voltage V_{GD}, drain current I_D, and temperature. Figure 3-10d shows a typical family of I_G-versus-V_{DG} characteristics, with I_D as a parameter. At a fixed value of I_D, I_G is a slowly varying function

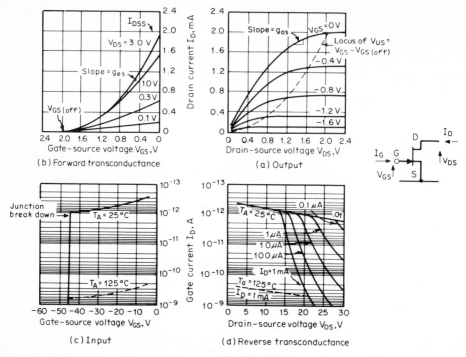

FIGURE 3-10   Characteristics of *n*-channel JFET.

of V_{DG} until the I_G breakpoint is reached. The breakpoint is a function of I_D. Beyond the breakpoint, I_G is an exponential function of V_{DG} and a linear function of I_D.

An understanding of the I_G-versus-V_{DG} characteristics is aided by considering the equivalent circuit shown in Fig. 3-11, which indicates that I_G is the sum of three components: I_1, I_2, and I_3. I_1 and I_2 are simple leakage currents of reverse-biased gate-source and gate-drain *p-n* junctions. They are the result of *thermal* ionization of carriers within the junction depletion layers and will approximately double in magnitude for each 10°C increase in junction temperature. I_1 and I_2 are approximately proportional to the square root of junction voltages V_{DG} and V_{SG}, respectively.

The I_3 component results from carriers generated within the drain-gate space-charge region due to *impact* ionization by the drain-current carriers. Thus I_3 is a *linear* function of I_D and an exponential function of V_{DG}. At low voltage the impact ionization rate is so low that the I_3 component is undetectable in the presence of the thermally generated components I_1 and I_2. Impact ionization rate, however, is an exponential

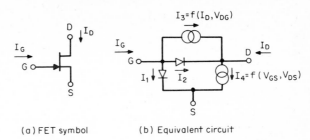

(a) FET symbol   (b) Equivalent circuit

FIGURE 3-11   Equivalent circuit of JFET to analyze gate current.

function of electric field. Thus increasing V_{DG} causes a rapid increase in the I_3 component.

Impact ionization is also a function of carrier mobility. Since mobility has a negative temperature coefficient, I_3 *decreases* with increasing temperature. The broken-line curve in Fig. 3-10 shows I_G versus V_{DG} at a temperature of 125°C. It can be seen that the breakpoint caused by I_3 occurs at a higher value of V_{DG}.

3-8-3   Practical FET Cascode Circuits

Operation at V_{DG} values below the I_G breakpoint is desirable for amplifier applications requiring very high input resistance. If a large input-voltage range is required, the circuits shown in Fig. 3-12b and d have been found to be useful. The circuits maintain a low V_{DG} on the *input* FET Q_1; thus, the breakpoint of Q_1 is not encountered, and the I_3 component is undetectable. Gate current of Q_2 increases as *its* I_G breakpoint is exceeded; however, it does not contribute to input current and remains well below 0.1 percent of I_D until avalanche breakdown is approached. The curves of Fig. 3-12 compare the I_G versus V_{DG} characteristics of the circuits shown.

Another advantage of the Cascode circuit is a reduction in input capacitance due to a decrease in the Miller effect. Figure 3-12 shows C_{in} versus V_{DG} for the amplifier circuits. The great reduction of C_{in} of the source follower of Fig. 3.12d is caused by the "bootstrapping" of the Q_1 drain to its source. Thus, both the drain and the source of Q_1 are forced to "follow" the gate, greatly reducing the effect of both C_{gd} and C_{gs} upon C_{in}.

Figure 3-13 shows the transconductance and static output characteristics of the Cascode circuit for comparison; the broken-line curves indicate the characteristics of the single device Q_1. At the selected operating point ($I_D = 10$ mA) the transconductance is approximately the same

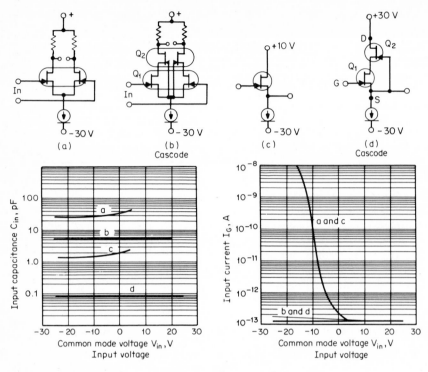

FIGURE 3-12   Cascode connections to reduce I_G and C_{in}.

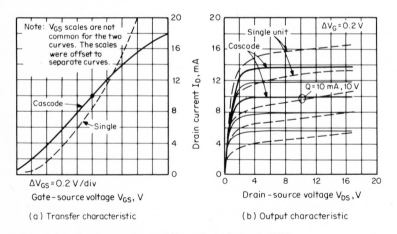

FIGURE 3-13   Comparison of Cascode and single FET.

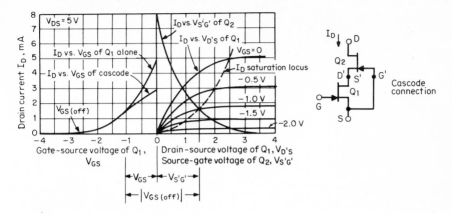

FIGURE 3-14   Characteristics of Cascode circuits.

as for the single device. The output conductance, however, is much lower for the Cascode connection.

Figure 3-14 will aid in understanding the Cascode characteristics. Note that V_{GS} of $Q_2 = V_{DS}$ of Q_1. The operating point for Q_1 will be at the intersection of the I_D-versus-V_{SG} curve of Q_2. The broken-line curve shows the approximate V_{DS} at which I_D saturation occurs for Q_1. The optimum operating point for Q_1 is near its I_D saturation. At lower values, transconductance begins to drop due to an increase in g_{ds}. Higher values will result in some increase in I_G.

The voltage at which drain current saturates is

$$V_{DS} = V_{GS} - V_{GS(off)} \tag{3-22}$$

$$V_{DG} = -V_{GS(off)} \tag{3-23}$$

To keep both Q_1 and Q_2 operating in drain-current saturation, the minimum drain-gate voltage of the Cascode connection should be

$$V_{DG(min)} = -V_{GS(off)(1)} - V_{GS(off)(2)} \tag{3-24}$$

If Q_1 and Q_2 are matched units, then

$$V_{DG(min)} = -2V_{GS(off)} \tag{3-25}$$

The Cascode FET pair, as shown in Fig. 3-15a, may be regarded as an active two-port. The midfrequency equivalent circuit shown in Fig. 3-15b may be reduced to the simplified circuit shown in Fig. 3-15c. The feedback capacitance C_{gd} and the output conductance g_{os} are reduced by the ratio g_{ds}/g_{fs}.

The voltage gain (A_V) of the *input* stage of the Cascode is approxi-

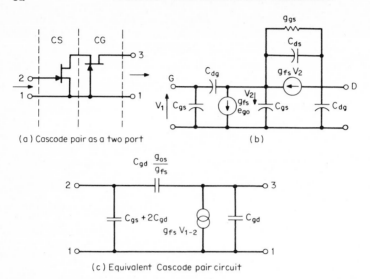

(a) Cascode pair as a two port   (b)

(c) Equivalent Cascode pair circuit

FIGURE 3-15   Cascode equivalent circuit.

mately -1; therefore, the input capacitance (including the Miller effect) is

$$C_{in} = C_{gs} + 2C_{gd} \qquad (3\text{-}26)$$

where C_{gd} and C_{gs} are values for the *input* FET.

Output capacitance is approximately C_{gd} of the common-gate output stage. Forward transconductance g_{fs} will be equal to the input-stage transconductance.

At a given drain current level the effective "bottoming" voltage (minimum V_{DG}) for the Cascode is about twice that for a single device; thus the power efficiency is lower. For small-signal amplifiers, which are the major application of the Cascode, this is not important. A commonly used Cascode configuration for FM and TV front ends is the dual-gate MOSFET, which in effect is an integrated Cascode pair (shown in Fig. 3-16). The 3N201 is an example of such a device. Its characteristics are discussed in Chapter 4.

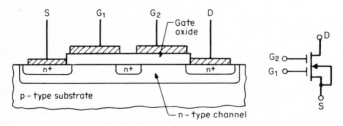

FIGURE 3-16   Dual-gate MOSFET.

The reduction of I_G and C_{in} by the use of the Cascode connection can be used to advantage in differential amplifiers. The common-mode voltage range can be greatly increased and the input bias current reduced. The Cascode differential amplifier is further discussed in Sec. 3-10.

3-9 SOURCE FOLLOWER (COMMON-DRAIN AMPLIFIER)

Too little knowledge of biasing methods for FET amplifiers sometimes keeps engineers from making maximum use of FETs in circuit designs. The common-drain amplifier, or source follower, is a particularly valuable configuration. Its high input impedance and low output impedance make it very useful for impedance transformations between FETs and bipolar transistors.

Figure 3-17 shows 10 circuits, which represent virtually every source-follower configuration. Understanding these circuits will help the designer obtain consistent circuit performance despite wide device variations.

There are two basic connections for source followers: with and without gate feedback. Each connection comes in several variations. In Fig. 3-17, circuits a through e have no gate feedback; their input resistances are approximately equal to R_G. Circuits f through k employ feedback to their gates to increase the input resistance above R_G.

Before getting into the details of bias-circuit design, note several general observations that can be made.

○ Circuits a, d, and f can accept only positive and small negative signals because these circuits have their source resistors connected to ground. A signal equal to $V_{GS(off)}$ will cut off I_D. The other circuits can handle large positive and negative signals limited only by the available supply voltages and device breakdown voltage.

○ Circuits c, d, e, h, j, and k employ current sources to improve drain-current (I_D) stability and increase gain.

○ Circuits d, e, and k employ FETs as current sources. In circuit d, Q_2 must have a lower cutoff voltage $V_{GS(off)}$ and a lower zero gate-voltage drain current I_{DSS} than Q_1.

○ Circuits e, g, h, and k employ a source resistor R_S which may be selected to set the quiescent output voltage equal to zero.

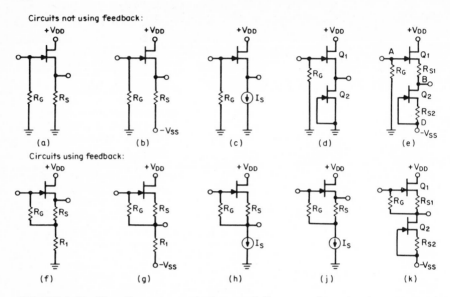

Circuits not using feedback:

(a)   (b)   (c)   (d)   (e)

Circuits using feedback:

(f)   (g)   (h)   (j)   (k)

FIGURE 3-17 Virtually every practical source-follower configuration is represented in this collection of 10 circuits. Configurations *(a)* through *(e)* do not employ gate feedback; the corresponding ones in the bottom row do. (Reprinted with permission from J. S. Sherwin, "Build Better Source Followers 10 Ways," *Electronic Design*, Vol. 18, No. 12, copyright Hayden Publishing Co., Inc., 1970.)

○ Circuits *e* and *k* use matched FETs. R_S is selected to set I_D near the specified low-drift operating current. The input-output offset is zero if $R_{S1} = R_{S2}$ and the FETs are matched.

3-9-1   Biasing without Feedback Is Simple

In Fig. 3-17, the no-feedback circuits *a* through *e* use simple biasing techniques. Circuit *a* is a self-bias configuration; the voltage drop across R_S biases the gate (which draws essentially zero current) through resistor R_G. Since no gate-to-source voltage V_{GS} can be developed when $I_D = 0$, the self-bias load line passes through the origin as shown in Fig. 3-18. For the 2N4339 FET, whose limiting transfer characteristics are used as an example, the quiescent drain current is seen to lie between about 0.25 and 0.55 mA when a 1-kΩ source resistor is used. The quiescent output voltage lies between +0.25 and +0.55 V.

Circuit 3-17*b* is another example of source-resistor biasing with a $-V_{SS}$ supply added. The advantage over circuit 3-17*a* is that the signal voltage can swing negative to approximately $-V_{SS}$. Two bias lines are shown, one for $V_{SS} = -15$ V and the other for $V_{SS} = -1.6$ V (see Fig.

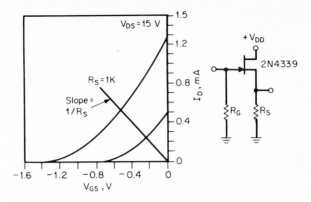

FIGURE 3-18 Self-biasing (Fig. 3-17*a*) uses the voltage dropped across the source resistor R_S to bias the gate. The load line passes through the origin and has a slope of $-1/R_S$. (Reprinted with permission from J. S. Sherwin, "Build Better Source Followers 10 Ways," *Electronic Design*, Vol. 18, No. 12, copyright Hayden Publishing Co., Inc., 1970.)

3-19). For the first case, the quiescent output voltage lies between +0.18 and +0.74 V. For the second, it lies between +0.3 and +0.82 V.

The bias load line for circuit 3-17*c* is just a horizontal line (I_D = constant). The quiescent output voltage is between +0.15 and 0.7 V for $I_D = 0.3$ mA.

Circuit 3-17*d* is similar to 3-17*c* except that the $V_{GS} = 0$ output characteristic of FET Q_2 is used as a current source. As seen in Fig. 3-20,

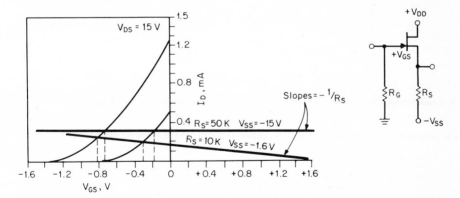

FIGURE 3-19 Adding a V_{SS} supply to the self-bias circuit (Fig. 3-18) allows it to handle large negative signals. The load line's intercept with the V_{GS} axis is at V_{SS}. Bias lines are shown for $V_{SS} = -15$ V and $V_{SS} = -1.6$ V. (Reprinted with permission from J. S. Sherwin, "Build Better Source Followers 10 Ways," *Electronic Design*, Vol. 18, No. 12, copyright Hayden Publishing Co., Inc., 1970.)

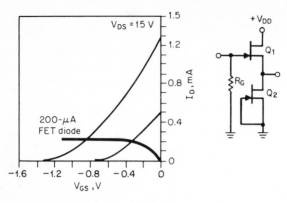

FIGURE 3-20 FET Q_2 does not behave like an ideal current source when its V_{DS} gets very small (Fig. 3-17d). Therefore, Q_1 should have a significantly larger $V_{GS(off)}$ than Q_2. (Reprinted with permission from J. S. Sherwin, "Build Better Source Followers 10 Ways," *Electronic Design*, Vol. 18, No. 12, copyright Hayden Publishing Co., Inc., 1970.)

Q_2 does not supply constant current when its V_{DS} gets very small. This technique should, therefore, be used only to bias FETs whose V_{GS} at the desired I_D is higher than the equivalent $V_{GS(off)}$ of the current-source FET diode.

A pair of matched FETs is used in the circuit of Fig. 3-17e, one as a source follower and the other as a current source. The operating drain current I_{DQ} is set by R_{S2}, as indicated by the load line of Fig. 3-21. The drain current may be anywhere from 0.20 to 0.42 mA, as shown by the limiting transfer characteristic intercepts; however, $V_{GS1} = V_{GS2}$ because the FETs are matched.

Since $I_{D1} = I_{D2}$ and $V_{GS1} = V_{GS2}$, choosing $R_{S1} = R_{S2}$ will ensure that the voltage from point A to B equals the voltage from point C to D (Fig. 3-17e). This source follower, therefore, exhibits zero or near-zero offset. If the FETs are temperature-matched at the operating I_D, the source follower will exhibit zero or near-zero temperature drift.

3-9-2 Biasing with Feedback Increases Z_{in}

Each of the feedback-type source followers (Fig. 3-17f through k) is biased by a method similar to that used with the nonfeedback circuit above it. However, in each case, R_G is returned to a point in the source circuit that provides almost unity positive feedback to the lower end of R_G. If R_S is chosen so that R_G is returned to zero dc volts (except in circuits 3-17f and j), then the input/output offset is zero. R_1 is usually much larger than R_S.

Circuit 3-17f is useful principally for ac-coupled circuits. R_S is usually much less than R_1 to provide near-unity feedback. The bias load line is set by R_S (Fig. 3-22). The output load line, however, is determined by the sum of $R_S + R_1$. The feedback voltage V_{FB}, measured at the junction of R_S and R_1, is determined by the intercept of the $R_S + R_1$ load line with the V_{GS} axis. The quiescent output voltage V_S is $V_{FB} - V_{GS}$.

In the circuit of Fig. 3-17g, R_S can be trimmed to provide zero offset. As the curves of Fig. 3-23 show, R_S will be between 670 Ω and 2.5 kΩ. R_S is much less than R_1. The source load line intercepts the V_{GS} axis at $V_{SS} = -15$ V.

Circuit 3-17h is almost the same as 3-17g; the difference is that resistor R_1 is replaced by a current source. Since an ideal current source has infinite impedance, the bias curve of circuit 3-17h differs from that of Fig. 3-17g (Fig. 3-24) in that the load line is perfectly flat.

Circuit 3-17j is similar to 3-17h except that the output is taken from the top of R_S to reduce the output impedance. R_S must be trimmed if the circuit is to work at all properly.

In Figure 3-24, the constant-current load line represents a 0.3-mA current source, and the effect of a 1-kΩ source resistor is shown. The offset voltage is seen to lie between 0.2 and 0.75 V. The intercept of

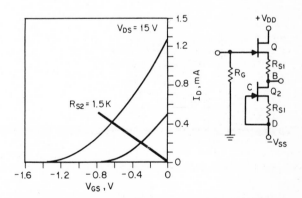

FIGURE 3-21   This load line is set by R_{S2} and Q_2, which acts as a current source (Fig. 3-17e). If its components are properly matched, the circuit will have zero or near-zero offset. (Reprinted with permission from J. S. Sherwin, "Build Better Source Followers 10 Ways," *Electronic Design*, Vol. 18, No. 12, copyright Hayden Publishing Co., Inc., 1970.)

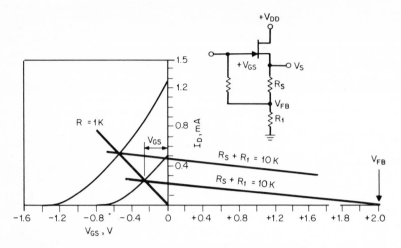

FIGURE 3-22 The bias load line is set by R_S, but the output load line is determined by $R_S + R_1$ when gate feedback is employed (Fig. 3-17*f*). The feedback V_{FB} is determined by the intercept of the $R_S + R_1$ load line and the V_{GS} axis ($V_{GS} \simeq V_{FB} - V_S$). (Reprinted with permission from J. S. Sherwin, "Build Better Source Followers 10 Ways," *Electronic Design*, Vol. 18, No. 12, copyright Hayden Publishing Co., Inc., 1970.)

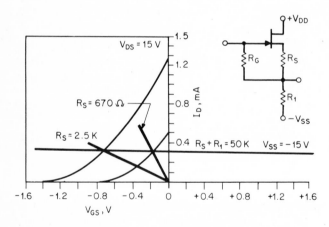

FIGURE 3-23 R_S can be trimmed to provide zero offset at some point between 670 Ω and 2.5 kΩ (Fig. 3-17*g*). The source load line intercepts the V_{GS} axis at $V_{SS} = -V_{GG} = -15$ V. Note that this load line is not perfectly flat. It has a slope of $-1/50$ K, because the current source is not perfect; it has a resistance of 50 kΩ. (Reprinted with permission from J. S. Sherwin, "Build Better Source Followers 10 Ways," *Electronic Design*, Vol. 18, No. 12, copyright Hayden Publishing Co., Inc., 1970.)

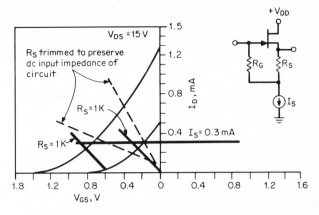

FIGURE 3-24 If R_S is not trimmed so that the load line passes through the origin, a voltage will appear at the gate, causing a reduction in dc input impedance. The incremental impedance will not be affected. (Reprinted with permission from J. S. Sherwin, "Build Better Source Followers 10 Ways," *Electronic Design*, Vol. 18, No. 12, copyright Hayden Publishing Co., Inc., 1970.)

the R_S load line and the V_{GS} axis sets the voltage at the junction of R_S and the current source (V_{FB}). For $R_S = 1$ kΩ, V_{FB} will be between -0.1 and $+0.45$ V. Since V_{FB} appears at the gate, it must be zero if the dc input impedance of the circuit is to be preserved. This can be done by trimming R_s, as shown dashed in Fig. 3-24. The biasing then becomes the same as for circuit 3-17*h*.

Biasing for circuit 3-17*k* is identical to that for circuit 3-17*e* except that feedback is added to raise the input impedance.

Figure 3-25 shows the general source-follower circuit, the equivalent small-signal, low-frequency equivalent circuit, and equations.

3-10   DIFFERENTIAL AMPLIFIER

A "differential amplifier" is one that provides an output proportional to the *difference* between the signals at the two input terminals while preventing an output occurring from a signal which is *common* to the two input terminals. The input terminals are "floating"—that is, neither input is grounded. The signal of interest is impressed between the two inputs. The output may be either a differential floating output or a single-ended output referenced to ground.

Good dc amplifier operation depends upon low-drift design. A single-

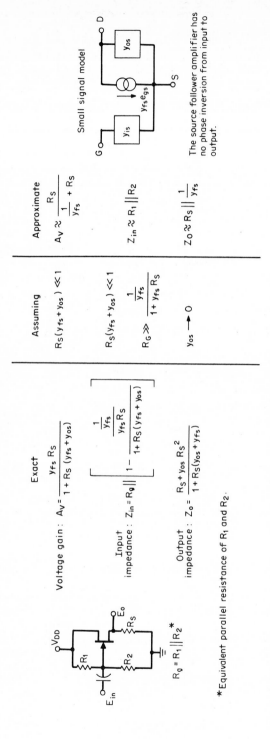

FIGURE 3-25 Summary of equations of source-follower amplifier (low-frequency model).

Small signal model

The source-follower amplifier has no phase inversion from input to output.

Exact

Voltage gain: $A_V = \dfrac{y_{fs}\, R_S}{1 + R_S\,(y_{fs} + y_{os})}$

Input impedance: $Z_{in} = R_g \left\| \dfrac{\dfrac{1}{y_{fs}}}{y_{fs}\, R_S}\left[1 - \dfrac{1}{1 + R_S\,(y_{fs} + y_{os})}\right] \right.$

Output impedance: $Z_o = \dfrac{R_S + y_{os}\, R_S^{\,2}}{1 + R_S\,(y_{os} + y_{fs})}$

*Equivalent parallel resistance of R_1 and R_2.

$R_g = R_1 \| R_2^{\,*}$

Assuming

$R_S(y_{fs} + y_{os}) \ll 1$

$R_S(y_{fs} + y_{os}) \ll 1$

$\dfrac{\dfrac{1}{y_{fs}}}{1 + y_{fs}\, R_S}$

$R_G \gg$

$y_{os} \longrightarrow 0$

Approximate

$A_V \approx \dfrac{R_S}{\dfrac{1}{y_{fs}} + R_S}$

$Z_{in} \approx R_1 \| R_2$

$Z_o \approx R_S \left\| \dfrac{1}{y_{fs}} \right.$

ended transistor stage may have drift which exceeds the magnitude of the signal to be amplified. Taking advantage of a pair of matched transistors and a design which provides common-mode rejection, the differential amplifier makes possible low-drift dc amplifiers.

3-10-1 Basic Design

Figure 3-26 shows a source-coupled pair of FETs. Assume that Q_1 and Q_2 are matched—that is, that they have identical transfer and output characteristics. With the input, e_i, at zero and e_{cm} at zero, the current source I_S will divide equally between Q_1 and Q_2, so that $I_{D1} = I_{D2}$ and $V_{D1} = V_{D2}$. The differential output voltage e_o will be zero. A change in the value of I_S or V_{DD} will not cause a differential output voltage. Now assume an input voltage e_i of +50 mV. This positive voltage tends to increase the current through Q_1. Any increase in I_{D1} must be offset by a corresponding *decrease* in I_{D2}, since the total current is fixed at I_S. If V_S were fixed, then ΔI_{D1} would be $e_i g_{fs}$; however, V_S is not fixed. Since ΔI_{D2} must equal minus ΔI_{D1}, v_{gs2} must equal minus v_{gs1} (assuming that g_{fs} of the two FETs is constant for small signals). Therefore, we can assume that

$$v_{gs1} = -v_{gs2} = \tfrac{1}{2}e_i \qquad (3\text{-}27)$$

(This assumes that $g_{fs1} = g_{fs2}$.)

The voltage changes at D_1 and at D_2 are

$$-v_{d1} = v_{gs1}g_{fs}R_D = \tfrac{1}{2}e_i g_{fs}R_D \qquad (3\text{-}28)$$

$$v_{d2} = v_{gs2}g_{fs}R_D = -\tfrac{1}{2}e_i g_{fs}R_D \qquad (3\text{-}29)$$

therefore
$$e_o = v_{d1} - v_{d2} = -e_i g_{fs}R_D \qquad (3\text{-}30)$$

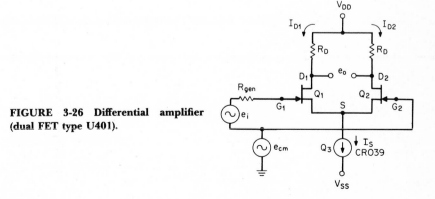

FIGURE 3-26 Differential amplifier (dual FET type U401).

Assume $g_{fs} = 10^{-3}$ mho and $R_D = 10$ kΩ. The output voltage with a 50-mV input signal is

$$e_o = -0.05 \times 10^{-3} \times 10,000 = -0.5 \text{ V}$$

The voltage amplification of this differential stage is

$$A_{\text{diff}} = \frac{e_o}{e_i} = -g_{fs}R_D \qquad (3\text{-}31)$$

It is the same as a single-stage common-source amplifier. [Output conductance g_{os} is neglected in Eqs. (3-28) through (3-31).]

A significant advantage of the differential amplifier is its ability to differentiate between common-mode input and differential input signals. Consider an input signal to be amplified e_i and an undesired signal e_{cm} to be rejected. The differential signal will result in an output voltage e_o given by Eq. (3-30). The common-mode signal, however, is applied to both gates equally, and if the devices are matched, there will be no differential change in drain currents, and hence there will be no differential output e_o.

The common source S of the two FETs will follow the common-mode voltage (offset by $-V_{GS}$); therefore V_{DS} will change by the magnitude of the common-mode voltage. V_{DS} should not be permitted to fall below the "pinchoff" region of the FET output characteristics. This condition will be met if $V_{DG} \geq -V_{GS(\text{off})}$; therefore, in the positive direction, the common-mode voltage should not exceed

$$\begin{aligned} e_{cm(\text{max})} &= V_D + V_{GS(\text{off})} \\ &= V_{DD} - I_D R_D + V_{GS(\text{off})} \end{aligned} \qquad (3\text{-}32)$$

(Note: $V_{GS(\text{off})}$ for the n-channel FET is a negative number.)

In the negative direction e_{cm} should not cause V_S to drop below the value required for the current source Q_3 to maintain its function of supplying adequate source current I_S. Other limits on e_{cm} are imposed by the maximum V_{DG} rating of the FETs and by the V_{DG} that causes I_G to become excessive.

The voltage-gain equation (3-31) shows that A_V is a linear function of R_D. For maximum gain a high value of R_D is desirable. Given a supply voltage, increasing the value of R_D decreases V_D, unless I_S is decreased to keep $R_D I_D$ constant. The choice of R_D is affected by:

1. Desired voltage gain A_V

2. Available supply voltage V_{DD}

3. Maximum common-mode input voltage $e_{cm(\text{max})}$

4. Maximum value of common-source current $I_{S(max)}$

5. Maximum differential input voltage e_i

6. High-frequency roll-off

We will go through some of the design steps for the circuit shown in Fig. 3-26. The following assumptions are made:

1. Common-mode voltage: $e_{cm} = \pm 8$ V.

2. Differential input voltage: $e_i = \pm 0.05$ V.

3. Voltage amplification: $A_V =$ as high as practical.

4. Available supply voltages: $V_{DD} = 15$ V $\pm 5\%$; $V_{SS} = -15$ V $\pm 5\%$. (Factors affecting the choice of the proper FET pair for Q_1 and Q_2 are discussed in Sec. 3-6. We will assume for this example that a general-purpose dual type has been selected.)

5. Q_1 plus Q_2 — dual FET type U401. (The matched characteristics of the U401 are specified at an operating I_D of 200 μA; therefore, we will select a current source Q_3 of approximately twice this value.)

6. Q_3 — Current regulator diode type CR039.

The maximum value of R_D is arrived at by solving for R_D in the equation

$$V_{DD(min)} - I_{D(max)}R_D = V_{D(min)} - V_{CM(max)} + V_{i(max)} + V_{DG(min)} \quad (3\text{-}33)$$

As a rule of thumb, to stay in the saturated current region of the FET output characteristics, V_{DG} should not fall below the value of $-V_{GS(off)(max)} = 2.5$ V; therefore, we set $V_{DG(min)} = 2.5$ V. $I_{D(max)}$ will occur when the differential input is at $e_{i(max)}$ and is given by

$$I_{D(max)} = \tfrac{1}{2}I_{S(max)} + \tfrac{1}{2}e_{i(max)}g_{fs(max)} \quad (3\text{-}34)$$

From the data sheets for the CR039 and the U401, $I_{S(max)} = 1.1 \times 0.390$ mA and $g_{fs(max)} = 1.6$ mmhos. Solving Eq. (3-33) for R_D yields

$$R_{D(max)} = \frac{V_{DD(min)} - V_{CM(max)} - e_{i(max)} - V_{DG(min)}}{\tfrac{1}{2}(I_{S(max)} + e_{i(max)}g_{fs(max)})}$$

$$R_{D(max)} = \frac{2(14.25 - 8 - 0.05 - 2.5)}{(1.1)(0.390)(10^{-3}) + (0.05)(1.6)(10^{-3})} \quad (3\text{-}35)$$

$$R_{D(max)} = 14{,}538 \ \Omega$$

We will use a value of 14,000 Ω for R_D and calculate the voltage gain using the data-sheet *minimum* value for g_{fs}.

$$A_V = -g_{fs}R_D = -(1,000)(10^{-6}) \times 14,000 = -14$$

We will now consider ways in which A_V may be increased, while still adhering to assumptions 1 through 4 above. Equation (3-31) shows that to increase A_V, the $g_{fs}R_D$ product must be increased. Other factors remaining constant, if R_D is increased, then $I_{D(max)}$ must be decreased to maintain the same minimum value for V_D. A decrease in I_D will result in a decrease in g_{fs}; however, as shown in Eq. (2-12) and Fig. 2-6, for a given FET the g_{fs} decrease will be proportional to the square root of the I_D decrease. Operating at a lower value of I_D, with a given device, *can* result in a higher voltage gain. In this example, if we reduce the value of I_S by a factor of 2, g_{fs} will be reduced by a factor of only $2^{1/2}$. Doubling R_D, then (to maintain the same I_DR_D), will increase A_V by a factor of $2^{1/2}$; thus the voltage gain would be increased from 14 to 19.8. Because of the doubling of R_D, the output resistance would be increased and the high-frequency corner would be decreased. Also, if the FETs are not operated at the data-sheet-specified value of I_D, the offset and drift specifications may not be met. Increasing the value of V_{DD} would also permit a higher value of R_D while maintaining the same $I_{D(max)}$. Care must be taken not to exceed the $V_{DG(max)}$ voltage rating of the FET.

3-10-2   Drift and Offset Compensation

JFET differential inputs are often used in dc amplifiers because they provide high input impedance and operate with extremely low input bias current. To achieve accurate measurements, the input offset and equivalent input drift must be made small with respect to the signal being measured. Methods of nulling or minimizing input offset and drift and how this compensation affects the common-mode rejection ratio (CMRR) will be examined.

An understanding of the effect of temperature changes upon the basic parameters of the FET aids in the application of temperature-induced drift compensation. For the purpose of this discussion, temperature drift is defined as the change in V_{GS} required to maintain a constant I_D as the temperature is changed. For a typical JFET, operating with a V_{DG} greater than $-V_p$, the relationship between I_D and V_{GS} was shown in Eq. (3-4).

As pointed out in Chapter 1, V_p is a function of the zero-bias depletion-layer width at the gate-channel p-n junction. The depletion width decreases with increasing temperature, which results in an increase in channel width and therefore an increase in $-V_p$. The temperature coefficient of $-V_p$ is about 2.2 mV/°C. This widening of the channel with increasing temperature tends to increase channel conduction; *however*, at the same time, increasing temperature causes a *decrease* in carrier mobility and thus in channel conductivity. Carrier mobility has a negative temperature coefficient of about 0.6%/°C. The positive channel-thickness coefficient and the negative mobility coefficient have opposite effects upon the temperature characteristic of I_D and I_{DSS}. In a thick-channel device (high $V_{GS(off)}$) the mobility factor dominates and I_{DSS} has a negative temperature coefficient. In a thin-channel unit the channel-thickness effect dominates and I_{DSS} has a positive temperature coefficient. At a particular channel thickness the temperature coefficients exactly compensate and I_{DSS} has a zero temperature coefficient. It has been shown that this particular thickness occurs in JFET devices having $V_{GS(off)}$ values of about -0.63 V. Units having *higher* $V_{GS(off)}$ values can be biased down to this critical channel dimension, so that I_D has zero temperature drift. Mathematically the zero-drift drain current I_{DZ} is

$$I_{DZ} = I_{DSS} \left(\frac{0.63}{V_{GS(off)}} \right)^2 \qquad (3\text{-}36)$$

Operating at drain currents other than I_{DZ} will result in a drift in V_{GS} as given by the equation

$$V_{GS(drift)} \approx 2.2 \text{ mV/°C} \left[1 - \left(\frac{I_D}{I_{DZ}} \right)^{1/2} \right] \qquad (3\text{-}37)$$

A plot of this characteristic is shown in Fig. 3-27. Differentiating Eq. (3-37) with respect to I_D produces a value K for the amount of drift change per unit change of I_D:

$$\frac{dV_{GS(drift)}}{dI_D} = \frac{(-1)(10^{-3})}{(I_D/I_{DZ})^{1/2}} = K \qquad \text{V/°C} \qquad (3\text{-}38)$$

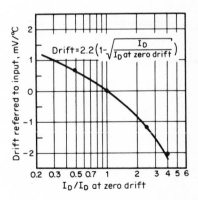

FIGURE 3-27 Drift of gate-source voltage in a single JFET drops with increasing normalized drain current. The slope of this curve is useful in determining the size of differential source resistance needed to null a given drift.

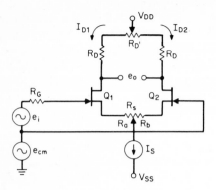

FIGURE 3-28 Differential amplifier with offset and drift compensation.

Ideally the differential amplifier produces an output only as a function of a differential input signal. If the temperature-induced drifts (dV_{GS}/dT) of the two FETs are equal, they appear to be a common-mode signal and do not produce an output. Unfortunately, few dual FETs have a perfect V_{GS} drift match; thus a differential offset will occur.

Since V_{GS} drift depends upon the value of drain current, the differential drift d may be reduced or eliminated by unbalancing the two drain currents by an amount ΔI_D:

$$\Delta I_D = \frac{d}{K} \tag{3-39}$$

where K is given by Eq. (3-38).

With the input voltage e_i held to zero, drain-current unbalance can be induced by adding unbalanced source resistors as shown in Fig. 3-28. A potentiometer R_S is used so that the unbalance can be adjusted. The resistor unbalance is a function of g_{fs}, I_S, and the required current unbalance.

$$\Delta I_D = I_{D2} = I_{D1} = \tfrac{1}{2}[g_{fs}I_S(R_a - R_b)] \tag{3-40}$$

Substituting Eq. (3-39) for ΔI_D and solving for R, the value of the R_S necessary to null a given drift becomes

$$R_S = R_a - R_b = \frac{2d}{g_{fs}I_S K} \tag{3-41}$$

Minimum values of g_{fs}, I, and I_S are used to ensure an adequate value for R_S.

Even if the FETs had no output initially, the introduction of the source resistance R_S would probably introduce an offset. This offset is due to the unbalanced current ΔI_D flowing through the balanced load resistors R_D. In this case, output balance can be achieved by unbalancing the load resistors R_D to compensate for the unbalance in I_D. The potentiometer R_D in the drain circuit of Fig. 3-28 serves this function. Since the output offset will be A_V times the input offset, the drain-offset compensation resistor will need to be A_V times the source-drift-compensating resistor R_S.

$$R_{D1} = A_V R_S = g_{fs} R_D R_S \qquad (3\text{-}42)$$

The actual value of R_D' should be greater than this so that any initial FET pair offset can be compensated for. The value that takes this into account is

$$R_D' = g_{fs} R_D \left[R_S + \frac{2(V_{GS1} - V_{GS2})}{I_S} \right] \qquad (3\text{-}43)$$

where $V_{GS1} - V_{GS2}$ = maximum specified offset of the FET pair. With the circuit shown in Fig. 3-28, first R_S would be adjusted for minimum drift with temperature, then R_D' would be adjusted for zero offset.

If the differential FET pair is used as the input stage of an operational amplifier (op amp), as shown in Fig. 3-29, the drift-adjusting procedure is modified. This occurs because with the feedback loop closed, the high-gain amplifier tends to force $V_{D1} - V_{D2}$ to zero. Thus to introduce an unbalance in drain current (ΔI_D), the drain potentiometer is first adjusted. This causes an output offset which is then eliminated by adjusting the source potentiometer R_S. Assuming the amplifier offset is zero, the calculation of the value of R_D' and R_S' is the same as for the single-stage amplifier. Compensation of the op amp offset and drift can be achieved by modifying the calculations for R_{D1} and R_S. The amplifier drift divided by the input stage gain $(g_{fs} R_D)$ should be added to the value d in Eq. (3-39).

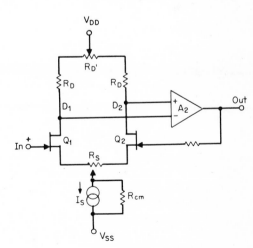

FIGURE 3-29 Differential FET input for operational amplifier.

The amplifier offset divided by the input stage gain should be added to the value $V_{GS1} - V_{GS2}$ in Eq. (3-43). Typically, this will change R'_D and R'_S by less than 10 percent.

Additional A_2 offset and drift will result from A_2 input-current offset and drift times the value of $R'_D + 2R_D$. Again, this is referred to the input by dividing by the voltage gain (A_V) of the input stage.

3-10-3   Common-Mode Errors

In the ideal differential amplifier a common-mode voltage e_{cm} would cause no differential output voltage e_o. The amplifier output would be a function only of the differential input e_i and would reject any common-mode input signal. In practice, this does not occur because nonideal conditions exist. One important figure of merit is a measure of the degree to which the differential amplifier rejects the common-mode signal in favor of the differential signal. This figure of merit is called the common-mode rejection ratio (CMRR).

Common-mode input signals cause two types of output errors in differential amplifiers. One type of error is a differential output signal caused by the common-mode input; the other is a common-mode output signal where both outputs change equally and in the same direction. Since the op amp which follows the FET input stage will typically amplify the differential-mode error 80 dB more than the common-mode error (assuming its CMRR is 80 dB), the differential-mode error is by far the more important of the two.

Differential-mode error is the result of unequal gain between the two sides of the differential amplifier. This can be caused by imprecise transconductance matching of the dual FETs, unbalanced drain resistances, unbalanced source resistances, imprecise output conductance matching of the FETs, or any combination of these factors. Thus, if the drain or source resistances are adjusted for drift or offset nulling, the differential output error will be increased and CMRR degraded.

Consider the case where the differential amplifier has slightly unbalanced gain. An equivalent circuit is shown in Fig. 3-30.

A figure of merit CMRR can be derived for differential output errors:

$$\text{CMRR}_{(\text{diff})} = \frac{A_{\text{diff}}}{A_{CM(\text{diff})}} \tag{3-44}$$

The differential-mode gain, assuming both halves of the differential amplifier are nearly equal, can be approximated as

$$A_{\text{diff}} \approx \frac{g_{fs}R_D}{1 + g_{fs}R_S}\left(\frac{1}{1 + g_{os}R_D}\right) \tag{3-45}$$

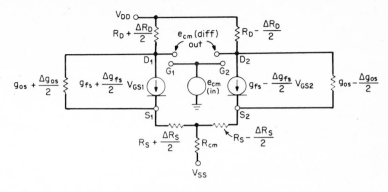

FIGURE 3-30 Differential output equivalent circuit.

for $1/g_{os} \gg R_S$. For $R_S \ll 1/g_{fs}$ and $R_D \ll 1/g_{os}$, this simplifies to the familiar expression

$$A_{\text{diff}} \approx g_{fs} R_D \qquad (3\text{-}46)$$

If the circuit is balanced except for the drain resistors, the differential-mode error resulting from a common-mode input signal becomes

$$A_{cm(\text{diff})} \approx \frac{\Delta R_D}{2 R_{cm}} \qquad (3\text{-}47)$$

so

$$\text{CMRR}_{\text{(dB)}} = 20 \log \left(\frac{2 R_{cm}}{\Delta R_D} g_{fs} R_D \right) \qquad (3\text{-}48)$$

Similarly, the error introduced by unbalanced transconductances is taken into account:

$$\text{CMRR}_{\text{(dB)}} = 20 \log 2 \, \Delta g_{fs} R_{cm} \qquad (3\text{-}49)$$

When unequal output conductances are taken into account,

$$\text{CMRR}_{\text{(dB)}} = 20 \log \frac{2 g_{fs} R_{cm} R_o}{\Delta R_o} \qquad (3\text{-}50)$$

and with unbalanced source resistors,

$$\text{CMRR}_{\text{(dB)}} = 20 \log \frac{2 R_{cm}}{\Delta R_S} \qquad (3\text{-}51)$$

assuming $1/g_{fs} \gg R_S$.

A more general equation, resulting from the combination of all the above factors, is

$\text{CMRR}_{(dB)} = 20 \log$

$$\left[\frac{2g_{fs}R_{cm}}{(\Delta R_S + 1/\Delta g_{fs})/(R_S + 1/g_{fs}) \pm \Delta R_D/R_D \pm \Delta R_o/R_o}\right] \quad (3\text{-}52)$$

The other type of error, common-mode output voltage, is important when only one of the outputs of the differential stage is used, or when the following stage has a poor CMRR. Figure 3-31 shows how common-mode output error can be measured. If common-mode gain for common-mode output errors is defined as

$$A_{CM(cm)} = \frac{V_{D1}}{V_{G1}} = \frac{V_{D2}}{V_{G2}} \quad (3\text{-}53)$$

then an expression may be found by noting that

$$V_D = -g_{fs}V_{GS}R_D \quad (3\text{-}54)$$

However,

$$V_{GS} = V_G - V_S \quad (3\text{-}55)$$

and

$$V_S = \frac{-2V_D R_{CM}}{R_D} \quad (3\text{-}56)$$

Thus,

$$V_D = -g_{fs}V_G R_D - g_{fs}2R_{CM}V_D \quad (3\text{-}57)$$

and

$$A_{CM(cm)} = \frac{V_D}{V_G} = \frac{-g_{fs}R_D}{1 + 2g_{fs}R_{CM}} \quad (3\text{-}58)$$

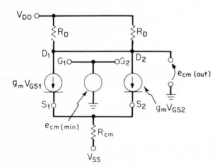

FIGURE 3-31 Common-mode output equivalent circuit.

It is common practice to assign a figure of merit to a differential amplifier. This figure of merit is "common-mode rejection ratio," usually abbreviated as $\text{CMRR}_{(cm)}$. $\text{CMRR}_{(cm)}$ is defined as

$$CMRR_{(cm)} = \frac{A_{diff}}{A_{CM(cm)}} \qquad (3\text{-}59)$$

When Eqs. (3-31) and (3-58) are substituted into Eq. (3-59), it is apparent that the common-mode rejection ratio for common-mode output errors is

$$CMRR_{(cm)} \approx \frac{-g_{fs}R_D}{-g_{fs}R_D/(1 + 2g_{fs}R_{CM})} = 1 + 2g_{fs}R_{CM} \qquad (3\text{-}60)$$

Most op amps have single-ended outputs. In such cases, there can be only one result of a common-mode input signal: an error in the output. In a FET input op amp the total CMRR will be a function of the common-mode and differential output errors of the preamplifier as well as the CMRR of the second stage. To calculate the combined CMRR of both stages of the FET input op amp, refer to Eq. (3-59), which is the definition of CMRR.

A two-stage amplifier is shown in Fig. 3-32. CMRR in this circuit is calculated as follows:

$$CMRR = \frac{A_{diff}}{A_{CM}} \qquad (3\text{-}61)$$

$$= \frac{(A_{diff1})(A_{diff2})}{(A_{CM(cm)})(A_{CM2}) + (A_{CM(diff)1})(A_{diff2})}$$

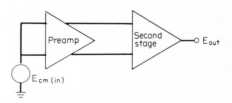

FIGURE 3-32 Two-stage op amp CMRR model.

Substituting Eq. (3-59) into Eq. (3-61), we have

$$CMRR_{total} = \frac{1}{1/(CMRR_{(cm)1})(CMRR_2) + 1/(CMRR_{(diff)1}} \qquad (3\text{-}62)$$

where $CMRR_{(cm)1}$ is the CMRR referred to in Eq. (3-60) for the FET preamplifier, $CMRR_2$ is the CMRR specification of the second stage, and $CMRR_{(diff)1}$ is the CMRR referred to in Eq. (3-52) for the FET preamplifier.

In a common-drain stage, the common-mode gain with respect to common-mode output error ($A_{CM(cm)}$) is unity (when $A_{diff} = 1$). Hence, the entire output common-mode signal is passed along to the second stage. All the equations in this section, except for Eqs. (3-53), (3-58), and (3-60), apply equally to common-drain preamplifiers.

3-10-4 Frequency Response

The frequency response of a FET differential amplifier is determined by two time constants, one for the input and one for the output.

The input time constant is formed by the generator impedance and the effective input capacitance C_{in}, where

$$C_{in} = \left(1 + \frac{g_{fs}R_D}{1 + g_{fs}R_S}\right)C_{DG} + \left(1 - \frac{g_{fs}R_S}{1 + g_{fs}R_S}\right)C_{GS} + C_{stray} \quad (3\text{-}63)$$

and

$$f_{in} = \frac{1}{2\pi C_{in} R_G} \quad (3\text{-}64)$$

The output time constant is formed by the drain resistance and the output capacitance C_{out}, where

$$f_{out} = \frac{1}{2\pi C_{out} R_D} \quad (3\text{-}65)$$

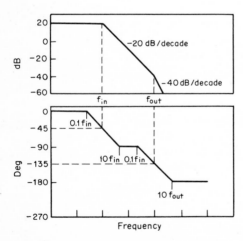

FIGURE 3-33 Phase and gain Bode plots for typical FET preamplifier.

and

$$C_{\text{out}} = C_{gd} + C_{\text{load}} + C_{\text{stray}} \qquad (3\text{-}66)$$

Equations (3-63) and (3-64) apply to common-drain preamplifiers as well as to common-source preamplifiers if $R_D = 0$.

For the case of the output time constant for the common-drain preamplifier,

$$f_{\text{out}} = \frac{g_{fs}}{2\pi C_{\text{out}}} \qquad (3\text{-}67)$$

where

$$C_{\text{out}} = C_{gs} + C_{sd} + C_{\text{load}} + C_{\text{stray}} \qquad (3\text{-}68)$$

Typically, the Bode plot of a FET preamplifier will resemble that shown in Fig. 3-33.

3-10-5   Stability and Phase Compensation

The stability criterion for any closed-loop system, including an operational amplifier, is that the phase shift around the loop must never reach 360° for any frequency at which the gain is unity or greater. Since the feedback around an op amp is in the inverting input, there is an intrinsic 180° phase shift. This dictates that the additional phase shift in the amplifier plus the phase shift of the feedback network must be less than 180° for all frequencies where gain is unity or greater.

Most commercially available operational amplifiers are compensated so that they have only 90° of phase shift where gain is unity or greater. This is demonstrated in Fig. 3-34. When a FET preamplifier is added ahead of an op amp, the total gain and phase response will be the sum of the response of the preamplifier and of the second stage. Often, this results in a phase shift of 180° or greater over much of the frequency range. An intolerable situation is thus created in that the total FET-input op amp will oscillate if the gain is set to a value where the phase shift is greater than 180°.

As a rule of thumb, an op amp will not oscillate if the closed-loop gain is such that the slope of the rolloff is no greater than 20 dB/decade. This point corresponds to a worst-case phase shift of 135°.

3-10-6   Selecting the Proper FET Pair

FET differential pairs can be broken down into four general types, according to their intended application. These types include low-leakage, low-noise, high-frequency, and general-purpose FET duals.

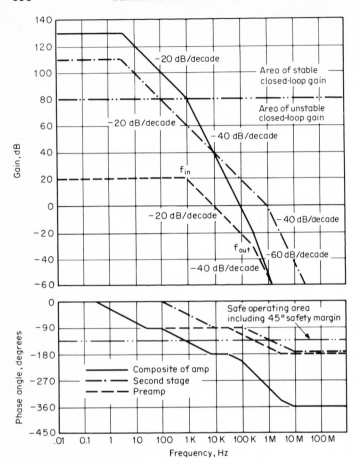

FIGURE 3-34 Compensated phase shift in op amps.

Low-leakage FETs generally have leakage currents in the range of 0.1 to 1.0 pA at 25°C. To achieve this low leakage, the active area of the device is made small. This small active area produces low g_{fs} and low capacitance. Although low-leakage FETs are preferred whenever low circuit leakage is the primary design criterion, the designer should consider using a general-purpose FET if slightly higher leakage can be tolerated. General-purpose devices offer better g_{fs}, offset, drift, and $\bar{e}_n$ than do low-leakage devices. One feature of the low-leakage FET which is not often specified in data sheets, but which is important to actual circuit performance, is the I_G breakpoint, that is, the drain-to-gate voltage at which the gate current rises rapidly. In practical circuits, the drain-to-gate voltage can reach 15 or 20 V, causing excessive leakage for FETs with low I_G breakpoints. If a common-mode voltage range in excess of 10 to 15 V is required, a Cascode input stage should be considered.

Low-noise FETs are designed primarily for low $\bar{e}_n$. They also have moderate g_{fs}, low g_{os}, and moderate leakage and breakdown voltage. Low-noise devices tend to have better drift and CMRR characteristics than do other types of dual FETs. A low-noise FET can produce the lowest noise operation of any FET when the generator impedance is below 1 to 10 MΩ. For higher generator impedances, low-leakage FETs may well provide the lowest overall noise performance of any transistor, bipolar or JFET, because of their lower noise current.

High-frequency dual FETs have very high g_{fs} and low capacitance. To achieve these characteristic leakage currents, breakdown voltage and the I_G breakpoint must be sacrificed. Because of these performance trade-offs, high-frequency FETs should be used only when their high-gain bandwidth is a design requirement. Dual high-frequency FETs are usually of hybrid (two-chip) construction rather than of a monolithic design because of the significantly lower capacitances between the two chips.

General-purpose dual devices are often the best choice for FET input op-amp applications, since they exhibit good g_{fs} and breakdown voltage and moderately low g_{os}, leakage current, and capacitance. General-purpose dual FETs also tend to have low $\bar{e}_n$ and good CMRR and drift characteristics.

Representative geometries of the four types of FETs just described are shown in Fig. 3-35 and provide a good insight into the relationship of device active area and performance characteristics.

When selecting any dual FET, two factors which affect the overall performance of the devices should be considered. One is the maximum value of $V_{GS(off)}$; a low maximum value of $V_{GS(off)}$ simplifies bias design and improves common-mode range, CMRR, offset, and drift. The other factor is the offset of the device. Although any value of offset can be nulled out of a circuit, the nulling process itself degrades CMRR and the drift performance of the circuit.

3-10-7 Selecting the Op-Amp Integrated Circuit

The monolithic IC op-amp portion of a FET input op-amp circuit either contributes to or solely determines five parameters of the complete circuit. For three of the parameters—offset, drift, and noise—the contributions are diminished in the complete op-amp circuit by the gain of the FET differential amplifier. In many circuits, however, these parameters are still significant and should not be ignored. Whenever possible, an IC op-amp device should be chosen which specifies maximum values

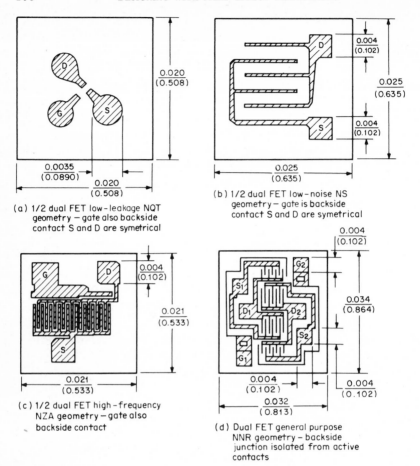

FIGURE 3-35 Four FET geometries. [All dimensions are given in inches (millimeters).]

for V_{os}, I_{os}, V_{drift}, I_{drift}, $\bar{e}_n$, and $\bar{i}_n$. Many IC op-amp data sheets will provide only typical values for these parameters. Typical values can be in error by an order of magnitude, and almost inevitably vary from lot to lot.

One parameter which is usually ignored on IC op-amp data sheets is so-called popcorn noise. Even though this parameter is missing from the data sheet, it is usually present in the op-amp IC and can be quite troublesome in low-noise designs. If low noise is a prime design criterion, an op amp specified for low popcorn noise should be selected.

Two other circuit parameters, output impedance and slew rate, are determined almost entirely by the IC op-amp portion of the circuit.

3-10-8 Design Example

This section presents a typical FET input op-amp design example (Fig. 3-36) using a Siliconix U401 n-channel junction FET. The U401 is designed to be operated with an I_D of 200 μA per device, or with an I_{CM} of 400 μA. A general-purpose operational amplifier should have a large common-mode range and good CMRR. These two requirements dictate that an active current source be used, such as the Siliconix CR043 current-regulator diode. The CR043 is ideal for this purpose, since it has a nominal current value of 430 μA $\pm$ 10% and a low temperature coefficient of less than 0.05%/°C typically, or 0.2 μA/°C.

Selection of drain resistor value involves a tradeoff between preamplifier gain and common-mode range. Common-mode range decreases and gain increases proportionally to increased value of the drain resistor. If a voltage drop of 3 V across the drain resistors is permissible, then the value of the drain resistors can be found by applying Ohm's law, as in

$$R_D = \frac{3 \text{ V}}{200 \ \mu\text{A}} = 15 \text{ k}\Omega$$

Gain of the preamplifier will be

$$A_{\text{diff}} \approx -g_{fs}R_D \tag{3-69}$$

From the U401 data sheet, g_{fs} at 200 μA operating current varies from a minimum of 1000 μmhos to a maximum of 1600 μmhos. For a worst-case design, the minimum value should be used:

$$A_{\text{diff(min)}} \approx (1 \times 10^{-3}) \cdot (1.5 \times 10^4) = 15.0$$

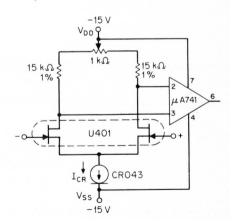

FIGURE 3-36 General-purpose design example circuit.

To keep the FETs in the current-saturated region of their output characteristics, V_{DG} should not fall below $-V_{GS(off)}$. The maximum value of V_G then will be

$$V_{G(max)} = V_{DD} - I_D R_D + V_{GS(off)} \qquad (3\text{-}70)$$

Two factors must be considered in determining the minimum value for V_G. One is the maximum value of V_{DG} determined by either the I_G breakpoint or the maximum V_{DG} rating. The other factor is the minimum voltage needed across the CR043 to maintain it in a high-impedance mode.

If $V_{DG(max)}$ is the limiting factor, then

$$V_{G(min)1} = V_{DD} - I_{D(min)} R_D - V_{DG(max)} \qquad (3\text{-}71)$$

If the minimum voltage $V_{CR(min)}$ across CR043 is the limiting factor, then

$$V_{G(min)2} = V_{SS} + V_{CR(min)} + V_{GS(min)} \qquad (3\text{-}72)$$

Both $V_{G(min)1}$ and $V_{G(min)2}$ should be computed and the most positive value used in the design. For our example the U401 data sheet $V_{DG} = 15$ V for the I_G specification, so that will be the value used for $V_{DG(max)}$. The CR043 data sheet indicates that the minimum "knee impedance" is measured at a V_F of 6 V, so that will be the value used for $V_{CR(min)}$.

The value of $V_{GS(min)}$ can be estimated by substituting the appropriate minimum and maximum values in the general FET V_{GS} equation:

$$V_{GS(min)} = V_{GS(off)(min)} \left[1 - \left(\frac{I_{D(max)}}{I_{DSS(min)}} \right)^{1/2} \right] \qquad (3\text{-}73)$$

From the U401 data sheet, $V_{GS(off)(min)}$ and $I_{DSS(min)}$ are -0.5 V and 0.5 mA. Maximum I_F of the CR043 is 473 μA; therefore $I_{D(max)}$ is $(\frac{1}{2})$ (473) $= 236.5$ μA. Using these values in Eq. (3-73), we get

$$V_{GS(min)} = (-0.5) \left[1 - \left(\frac{0.2365}{0.5} \right)^{1/2} \right] = -0.156 \text{ V}$$

Then with Eqs. (3-70), (3-71), and (3-72) we get

$$V_{G(max)} = 15 - (0.2365)(15) + (-2.5) = 8.95 \text{ V}$$

Minimum I_F of the CR043 $= 0.387$ mA; therefore $I_{D(min)} = 0.1935$ mA.

$$V_{G(min)1} = 15 - (0.1935)(15) - 15 = -2.9 \text{ V}$$

$$V_{G(min)2} = -15 + 6 + (-0.156) = -9.156 \text{ V}$$

If the common-mode input drops below -2.9 V, I_G may become excessive because V_{GD} will exceed 15 V. The common-mode voltage range for the amplifier then is from -2.9 to $+8.95$ V, if the I_G specification is not to be exceeded. If a higher I_G can be tolerated, then the common-mode range can go to -9.156 V.

The offset will be

$$E_{Tin} \approx E_{os1} + \frac{2I_{os}}{g_{fs}} + \frac{E_{os2}}{g_{fs}R_D} \tag{3-74}$$

From the U401 data sheet, $E_{os1} = 5$ mV, and $g_{fs(min)}$ at $I_D = 200$ μA is 1000 μmhos. From the μA741 op-amp data sheet, $I_{os(max)}$ is given as 200 μA. Therefore

$$E_{Tin(max)} = 5 \times 10^{-3} + \frac{2(2 \times 10^{-7})}{1 \times 10^{-3}} + \frac{5 \times 10^{-3}}{15.0} \approx 5.73 \text{ mV}$$

If both sides of Eq. (3-74) are divided by ΔT, then

$$\frac{E_{Tin}}{\Delta T} \approx 10 \times 10^{-6} + \frac{2(6 \times 10^{-10})}{9 \times 10^{-4}} + \frac{1 \times 10^{-5}}{15.0}$$

or

$$\frac{\Delta E_{Tin}}{\Delta T} \approx 12 \ \mu\text{V}/°\text{C} \qquad \text{worse case, assuming that the drain resistors are perfectly balanced}$$

For drain nulling, the value of the null potentiometer should be

$$R_N = \frac{2E_{Tin(max)}g_{fs(max)}R_D}{I_{CM(min)}} \tag{3-75}$$

or

$$R_N = \frac{2(5.8 \times 10^{-3})(1.6 \times 10^{-3})(1.5 \times 10^4)}{3.87 \times 10^{-4}}$$

or

$$R_N = 720 \ \Omega$$

However, to this approximation for R_N must be added the maximum drain resistor unbalance, which is 2 percent R_D for 1 percent tolerance resistors, or 300 Ω. Thus $R_N \approx 1$ kΩ. The output frequency of the preamplifier will be

$$f_{out} = \frac{1}{2\pi R_D C_{out}} \tag{3-76}$$

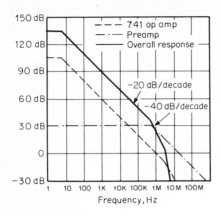

FIGURE 3-37   Bode plot of preamplifier and second stage.

where

$$C_{\text{out}} = C_{dg} + C_{\text{load}} + C_{\text{stray}} \tag{3-77}$$

or

$$C_{\text{out}} \approx 5 \text{ pF} + 2 \text{ pF} + 5 \text{ pF} \approx 12 \text{ pF}$$

and thus

$$f_{\text{out}} = \frac{1}{(6.28)(1.5 \times 10^4)(1.2 \times 10^{-11})} = 880 \text{ kHz}$$

The combined Bode plot of the preamplifier and the second stage is shown in Fig. 3-37. The Bode plot indicates that the op amp will be stable for any closed-loop gain of greater than 35 dB. For closed-loop gains of less than this value, one of the numerous forms of op-amp compensation must be used.

3-10-9   Cascode Differential Amplifiers

The common-mode voltage range of a FET differential amplifier can be increased by utilizing the Cascode configuration. Consider the differential stage shown in Fig. 3-38. The dual FET type U421 was chosen because of its low gate operating current I_G, which is specified as 0.1 pA maximum at $V_{DG} = 10$ V and $I_D = 30$ μA. An examination of the typical performance curves for the U421 reveals that the I_G breakpoint for this device is about 10 V V_{DG}, as shown in Fig. 3-39. This means that even with ideal bias conditions the common-mode voltage range cannot exceed 10 V without an appreciable increase in I_G occurring. Actually the maximum specified value of $-V_{GS\,(\text{off})}$ must be subtracted from the 10 V, leaving a common-mode range of 8 V. A further reduction

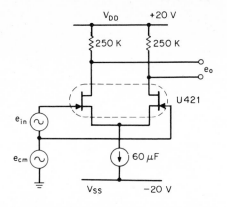

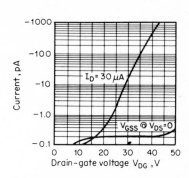

FIGURE 3-38   FET differential amplifier.

FIGURE 3-39   Leakage currents vs. drain-gate voltage.

must be allowed for $(1 - A_V)/2$ times the maximum differential signal. Thus, even though the U421 has a maximum gate-drain and gate-source voltage rating of -40 V, operating gate current will start to increase appreciably when V_{DG} exceeds about 10 V.

Common-mode voltage range, common-mode input resistance, and common-mode rejection ratio can be improved by the addition of another pair of FETs connected Cascode, as shown in Fig. 3-40. In the example, we utilize the Siliconix type U401 for the common-gate pair and still use the U421 for the common-source input pair. From a study of the characteristic curves for the U421, we conclude that a minimum

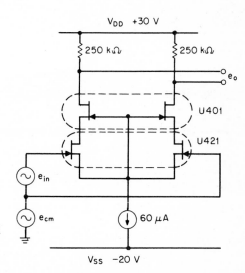

FIGURE 3-40   Cascode differential amplifier.

V_{DS} of 1 V should be allowed to ensure that it operates in drain-current saturation. This means that the minimum V_{GS} for the U401 at the operating I_D of 30 μA should be 1 V, and therefore a selection may be called for. Since the maximum $V_{GS(\text{off})}$ for the U401 is specified as 2.5 V, the range of V_{DS} for the U421 will be 1 V to slightly less than 2.5 V. V_{GS} at $I_D = 30$ μA is specified as -1.8 V maximum for the U421; therefore, the maximum V_{DG} of the U421 for this circuit will be about 4.3 V. This is well below the I_G breakpoint. The common-mode voltage range is now limited by the breakdown rating of the U401, which is specified as 50 V. Although 50 V exceeds the I_G breakpoint of the U401, its I_G is still typically less than 10 nA; since this current is supplied by the common-source current supply, it does not contribute to input leakage. The input bias current is virtually independent of the common-mode voltage because the V_{DG} of the *input* FETs does not change with common-mode voltage.

3-11   DISTORTION IN FET AMPLIFIERS

Harmonic distortion in any amplifier stage is principally the result of a nonlinear input-output transfer characteristic plus the effect of nonlinear loading on the signal source. As the transfer curves of most amplifying devices are nonlinear, a certain amount of distortion will be generated by any amplifier, whether it employs a vacuum tube, transistor, or FET. Linearity is affected by bias, operating voltage, load impedance, signal level, and device characteristics. Input-impedance variations may also have an effect under certain operating conditions. Distortion may be minimized in any application by operating the device under the most advantageous bias and load conditions.

3-11-1   Mathematical Analysis

As certain types of distortion are more detrimental than others in any given application, an understanding of the types and causes of distortion is desirable. The type and amount of distortion observed is a function of the extent to which the transfer curve departs from a straight line. Neglecting the effects of supply voltage (except for the vacuum-tube triode) and assuming neither load-current cutoff nor saturation, transfer curves of the several amplifier devices are closely approximated as follows:

1. Triode (vacuum tube)

$$I_P = K_1 \left(V_G + \frac{V_p}{\mu} \right)^{3/2} \qquad g_m = \frac{3}{2} K_1 \left(V_G + \frac{V_p}{\mu} \right)^{1/2} \qquad (3\text{-}78)$$

2. Pentode (vacuum tube)

$$I_P = K_2 \left(V_G + \frac{V_{G2}}{\mu_{G2}} \right)^{3/2} \qquad g_m = \frac{3}{2} K_2 \left(V_G + \frac{V_{G2}}{\mu_{G2}} \right)^{1/2} \qquad (3\text{-}79)$$

3. FET

$$I_D = K_3 \left(1 - \frac{V_{GS}}{V_p} \right)^2 \qquad g_m = \frac{2 K_3}{V_p} \left(1 - \frac{V_{GS}}{V_p} \right) \qquad (3\text{-}80)$$

4. Bipolar transistor

$$I_C = K_4 \left(e^{\lambda V_{BE}} - 1 \right) \qquad g_m = K_4 \lambda e^{\lambda V_{BE}} \qquad (3\text{-}81)$$

At the time of this writing, VMOS power FETs are available with a very linear region in their transfer characteristic. Although not designed for low-power small-signal amplifiers, they can be used for making low-distortion amplifiers if operated with a bias current in excess of 400 mA. This type 2N6659, for example, has a transfer curve of approximately

5. VMOS $\qquad I_D \simeq K_5 \left(V_{GS} - V_{TH'} \right) \qquad g_m = K_5 \qquad (3\text{-}82)$

K_5 for the 2N6659 is approximately 0.25 mho at values of I_D in the 400-mA to 1-A range. Below 400 mA the 2N6659 tends to follow Eq. (3-80).

To determine the degree of distortion, the transfer curves may be expressed as a power series where each term individually represents the fundamental or a distortion term. For example, the small-signal transfer function evaluated at a given set of dc operating conditions may be written as

$$i_p = g_m e_g + \frac{1}{2!} \frac{\delta g_m}{\delta V_G} e_g^2 + \frac{1}{3!} \frac{\delta^2 g_m}{\delta V_G^2} e_g^3 + \cdots + \frac{1}{n!} \frac{\delta^{n-1} g_m}{\delta V_G^{n-1}} e_g^n \qquad (3\text{-}83)$$

Note that each term includes a derivative of the transconductance term g_m. The first term represents the desired fundamental. The second represents a constant plus second harmonic, and the third represents the first and third harmonics. The first harmonic component of the third term adds to the fundamental, causing a loss in proportionality between input and output. The third-harmonic component causes cross-modulation and intermodulation distortion when two or more signals are present at the input.

With FETs, the second- and higher-order derivatives of g_m are zero; therefore, only a second harmonic is present and cross-modulation products are extremely low. In the case of vacuum tubes, fourth- and higher-order terms are negligible. However, the series does not converge so rapidly for the transistor; hence fourth- and higher-order harmonic distortion may be significant, and cross-modulation is more serious than with the vacuum tube.

As the second term of the series is proportional to curvature in the transfer characteristic, second-harmonic distortion may be minimized by operating on the most linear portion of the transfer curve.

The third term is proportional to rate of change of curvature, therefore third-harmonic and cross-modulation may be minimized in the same manner.

Although the FET promises to generate only second-harmonic distortion, the operating point must be carefully controlled for minimum distortion. Calculations made from an idealized or measured transfer curve may not necessarily tell the entire story. For example, the dc transfer curve of Fig. 3-1, normally plotted at a constant V_{DS}, does not take into account the effect of V_{DS} varying with signal or of g_{os} varying with I_D. Figure 3-5 shows both of these effects.

3-11-2   Sources of Distortion

Figure 3-41 shows a FET amplifier stage. Variations in e_{sig}, V_{GS}, R_L, $V_{GS(off)}$, g_{os}, and g_{fs} will have an effect on distortion. For low distortion, it is desirable to operate with small input signals near zero bias. Both R_L and V_{DD} should be high, and V_{DS} should be sufficiently above the knee of the output-characteristic curves but no greater than necessary. Also avoid the region near drain-to-gate breakdown to prevent the flow of gate current on signal peaks. Furthermore, certain FET geometries are preferred over others.

3-11-3   Transfer Curve and Output Conductance

Follow the instantaneous operating point up and left along the load line in Fig. 3-42. Note that g_{fs} increases as V_{GS} approaches zero. This causes second-harmonic distortion due to curvature of the transfer curve. What is not so apparent is that nonlinearities exist due to decreasing g_{os}, as I_D decreases. This effect is visible in the stylized sets of constant

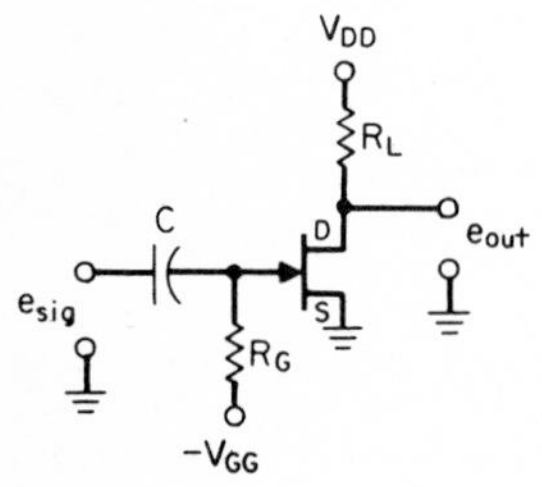

FIGURE 3-41   *RC amplifier.*

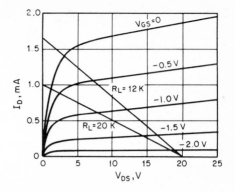

FIGURE 3-42 FET output characteristics.

g_{fs} output characteristics of Fig. 3-43. Set a exhibits constant g_{fs} and constant g_{os}, and set b exhibits a slight nonlinear change in g_{os} with I_D. Set b indicates an increase in gain as the operating point moves down and right along the load line. Reference to the gain equation indicates that the effect of decreasing g_{os} is less for large G_L, as expected from graphical analysis of Fig. 3-42. g_{fs} decreases with increasing V_{GS}. The decrease in g_{os} tends to offset the distortion produced by the nonlinear transfer curve. Thus at a certain operating point the two distortion sources nearly cancel to produce a point of minimum distortion.

$$\frac{e_o}{e_i} = \frac{g_{fs}}{1/R_L + g_{os}} \tag{3-84}$$

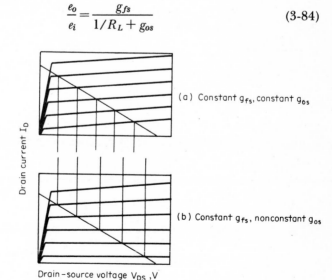

FIGURE 3-43 Idealized FET output characteristics: *(a)* **constant g_{fs}, constant g_{os}; *(b)* constant g_{fs}, nonconstant g_{os}.**

Expanding the power series transfer function of Eq. (3-83) for the FET gives the expression

$$i_d = \frac{2I_{DSS}}{V_p^2}(V_p - V_{GS})E_a \sin \omega t - \frac{I_{DSS}}{V_p^2}E_a^2 \left(\frac{\cos 2\omega t}{2}\right) \cdots \quad (3\text{-}85)$$

where E_a = peak signal amplitude.

Taking the ratio of the amplitude of the second harmonic to the fundamental, the following expression for percent second-harmonic distortion is derived:

$$\% \text{ 2d harmonic distortion} = \frac{25e_{\text{sig(pk)}}}{V_p - V_{GS}} \quad (3\text{-}86)$$

This expression is valid for small-signal distortion due only to transfer characteristic curvature. Required conditions are that the drain-voltage saturation region is avoided and that drain current flows during all portions of the cycle. Measurements of FET distortion confirm the expected results and, indeed, indicate close agreement with the distortion as calculated from Eq. (3-23).

3-11-4 Gate Bias and Signal Level

According to Eq. (3-86), small-signal distortion is directly proportional to signal level. The plot of total harmonic distortion vs. input signal of Fig. 3-44 provides verification, indicating that distortion increases linearly with input signal level. Although Fig. 3-44 was generated from data taken at zero gate bias, the distortion level is unaffected by the slight forward gate biasing on negative signal peaks because the gate input impedance stays high for low values of forward bias.

Input circuit distortion does not become measurable until the rms input signal level exceeds 100 mV. This is borne out by Fig. 3-45, which plots input distortion vs. input signal for gate-circuit time constants of 10 to 1000 ms. The reason for increasing distortion with input signal

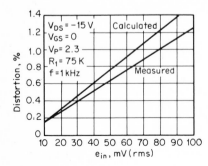

FIGURE 3-44 Distortion vs. input signal level.

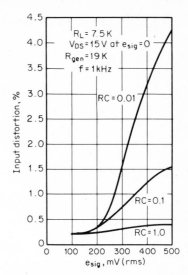

FIGURE 3-45   Distortion vs. input voltage and time constant.

is that the gate draws current from the signal source on input peaks; distortion is reduced by employing the equivalent of grid-leak bias. In this manner, the direct current drawn through the gate-return resistor develops a gate bias approximately equal to the peak forward signal voltage. So long as the input capacitor remains charged, the gate ceases to draw current on signal peaks. Thus distortion becomes a function of frequency, e.g., reducing the frequency to 100 Hz will have the same effect as reducing $T = RC$ by a factor of 10. Figure 3-45 indicates that for low distortion when the gate conducts on signal peaks, RC must be about 1000 times the period of the lowest frequency to be handled. If the generator impedance is quite low, say a few hundred ohms, the input distortion will be reduced considerably. However, as this is not a normal operating condition, the gate should be operated at a bias level sufficient to ensure that signal peaks will not forward-bias the gate by more than 100 to 200 mV.

Further reference to Eq. (3-86) indicates that distortion increases as V_{GS} approaches V_p, or that for $V_{GS} = 0$, distortion is inversely proportional to V_p. Distortion versus V_{GS} is plotted in Fig. 3-46 for two values of input signal for a device with $V_{GS(off)} = 2.3$ V. Figure 3-47 plots distortion against V_P for $V_{GS} = 0$. Rather close agreement with calculated values in evident in both plots.

3-11-5   Drain Voltage

The effects of V_{DS} and g_{os} on distortion are illustrated in Fig. 3-48. The V_{DS} effect is in reality a g_{os} effect, as may be seen from reference to

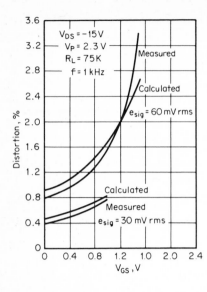

FIGURE 3-46   Distortion versus V_{GS}.

the g_{os}-versus-V_{DS} curves of Fig. 3-49. Distortion due to g_{os} becomes large at low V_{DS} due to the very high value of g_{os}, while the distortion at high V_{DS} is principally due to the variation in g_{fs}. As V_{DS} is decreased, the g_{os}-induced distortion increases and, as previously pointed out, acts to counteract the g_{fs}-induced distortion. At a specific and fairly low V_{DS} the two out-of-phase distortions nearly cancel to produce a minimum in the distortion curve. Operation at the point of minimum distortion .is not particularly recommended, since its location is uncertain and very near a region of excessive distortion.

Operation below the point of minimum distortion at low V_{DS} results in a significant increase in nonlinear distortion due to drain saturation,

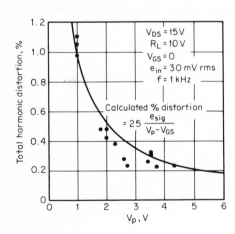

FIGURE 3-47   Distortion versus V_P.

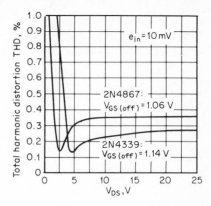

FIGURE 3-48 Distortion versus g_{os}.

reflected in a rapidly increasing g_{os}. This effect is also visible in Fig. 3-42 as the operating point moves up and to the left along the load line approaching the knee of the output curves.

3-11-6 FET Geometry

Physical geometry of the FET has an effect on distortion, as shown in Fig. 3-48, where two devices of approximately equal $V_{GS(off)}$ generate significantly different distortion levels. These distortion levels are due to different gate lengths, which affect the value of g_{os} and the rapidity with which device output characteristics shift from triode to pentode type as V_{DS} increases. Figures 3-50 and 3-51 show the output characteristics of the two devices. The differing values of g_{os} are apparent.

The 2N4867 is a very long-channel device and exhibits the lowest value of g_{os}. The I_D transition from unsaturated to saturated is quite rapid, as shown in Fig. 3-50. Distortion is high because the low g_{os} only slightly reduces that induced by g_{fs}. On the other hand, distortion at

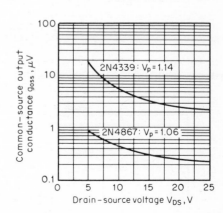

FIGURE 3-49 g_{os} versus V_{DS}.

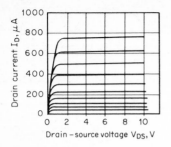

FIGURE 3-50 2N4867 output characteristics.

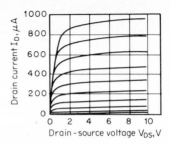

FIGURE 3-51 2N4339 output characteristics.

low V_{DS} is low. Where drain-supply voltages are limited, the long-gate device is preferred over other FET types because the operating point may be closer to the knee of the output curve.

The 2N4339 is a short-channel device exhibiting rather high g_{os} and a poorly defined transition region from high g_{os} to low g_{os}, as shown in Fig. 3-52. The high g_{os} results in a fairly low distortion so long as adequate drain voltage is available. The short-gate device is probably the best choice where minimum distortion is a requirement if supply voltage is not severely limited.

3-11-7 Load Resistance

The load resistance also affects distortion level. The effect is again a function of g_{os}. Figure 3-52 shows how distortion decreases with increasing R_L. The improvement is greater for short- than for long-channel devices, as g_{os} is greater for the former. The reason for an improvement with increasing R_L is that the g_{os}-induced distortion has a greater opportunity to counteract g_{fs} distortion as the load line becomes more nearly horizontal. Consider, for example, that g_{os} effects are zero when $R_L = 0$ and the load line is vertical. Conversely, g_{fs} effects are near zero when $R_L = \infty$ and the load line is horizontal. At some point with high R_L, a distortion minimum will occur beyond which g_{os} distortion is greater than g_{fs} distortion.

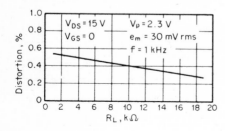

FIGURE 3-52 Distortion vs. load.

3-11-8 Rules for Low Distortion

From the preceding discussion, it follows that FET amplifiers may be operated at very low levels of distortion. However, close attention must be given to the dc operating point, bias level, load resistance, and FET characteristics. The rules to follow, listed in descending order of importance, are:

1. Maintain V_{DS} high enough that peak output signal swing will not reduce V_{DS} below 2 to 4 times $-V_{GS(off)}$.

2. Maintain V_{GS} at such a point that peak input signal swing will not forward-bias the gate junction by more than 200 mV.

3. Do not operate near drain-gate breakdown voltage unless utilizing a low signal source impedance.

4. Maintain a minimum V_{GS} consistent with other circuit requirements. More properly, maximize $V_{GS} - V_{GS}$.

5. Minimize input signal level.

6. Use a high value of load impedance.

7. Maintain V_{DS} at the lowest value consistent with rules 1 and 3.

Note that rules 4 and 6 indicate operation at the highest practical gain commensurate with power-supply and frequency-response limitations.

A particularly practical viewpoint is to consider distortion as a percentage of output voltage rather than of signal voltage as represented by Eq. (3-86). The combined effects of FET parameters $V_{GS(off)}$, I_{DSS}, g_m, and stage gain may then be determined.

Consider the stage gain expressed in Eq. (3-84) simplified to

$$\frac{e_o}{e_i} = g_m R_L \tag{3-87}$$

Combine this with Eq. (3-86) to obtain distortion in terms of output voltage.

$$\% \text{ distortion} = \frac{25 e_o(pk)}{g_m R_L (V_p - V_{GS})} \tag{3-88}$$

If R_L is related to supply voltage and FET characteristics as follows:

$$R_L = \frac{V_{DD} - 2V_p}{2 I_D} \tag{3-89}$$

For maximum output signal swing, the FET characteristic equations

$$I_D = \frac{I_{DSS}}{V_p^2} (V_p - V_{GS})^2 \tag{3-90}$$

$$g_m = \frac{2 I_{DSS}}{V_p^2} (V_p - V_{GS}) \tag{3-91}$$

may be used with Eqs. (3-88) and (3-89) to find output distortion.

$$\% \text{ distortion} = \frac{25 e_{o(pk)}}{V_{DD} - 2 V_p} \tag{3-92}$$

Equation (3-92) shows that FET amplifier distortion for a given output voltage is independent of I_{DSS} and g_{fs}. Rules 4 and 5 above become of little importance for the case where e_{sig} is small and R_L may be increased without regard to bandwidth or other design considerations. High V_{DD} and low $V_{GS(off)}$ allow high R_L, high gain, and thus low e_{sig} to produce a given e_o. Thus distortion is minimized due not only to the low signal required, but also to the advantages of high R_L already discussed but not expressed in Eq. (3-92).

For lowest distortion the FET selection will depend upon the available input signal, supply voltage, and required bandwidth. For large input signal and large bandwidth, a high $V_{GS(off)}$ is desired unless source degeneration is applied. For small input signal or low V_{DD}, a low $V_{GS(off)}$ unit is desired. The choice will depend upon the specific conditions of application. In any case, a FET exhibiting high g_{oss} is desired. This characteristic may be evaluated by reference to the output characteristic curves for any particular FET. Output characteristic curves with large slope mean high g_{oss}.

Note that no attempt has been made to play off one characteristic against another. For example, a low $V_{GS(off)}$ unit from any one given geometry will exhibit a slightly higher g_{fs} at a given drain current than will the high $V_{GS(off)}$ unit of the same family. This means that the device may be operated at low I_D for higher gain g_{fs} than is attainable with the high V_p unit. Few of these tradeoffs can improve the distortion level.

3-12   AUDIO-FREQUENCY NOISE CHARACTERISTICS

The purpose of this section is to identify and characterize audio-frequency noise in junction field-effect transistors. Emphasis is placed on basic device characteristics rather than on end applications, since it is important for the circuit designer to know the salient noise behavior of the FET, and how those characteristics may be specified by production-oriented test parameters.

3-12-1 Defining FET Noise Figure

For analysis, it is convenient to represent noise in a FET by assuming that an ideal noise-free device has two external noise sources, $\bar{e}_n$ and $\bar{i}_n$. These noise sources are chosen to have the same output as would an actual noisy FET. An equivalent circuit is shown in Fig. 3-53.

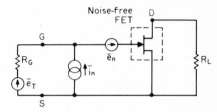

FIGURE 3-53 Representing noise in an ideal FET.

A noise factor F is a figure of merit of a device with respect to the resistance of a generator. To calculate a noise factor, a source resistor R_G, with a thermal noise voltage $\bar{e}_T$, is added to the circuit. A noise factor F may be defined as

$$F = 1 + \frac{\text{noise power of FET referred to input}}{\text{noise power due to } R_G} \tag{3-93}$$

The thermal noise voltage across R_G is

$$\bar{e}_T = (4kTR_GB)^{1/2} \tag{3-94}$$

where $k = 1.380 \times 10^{-23}$ J/K (Boltzmann's constant)
$T =$ temperature in Kelvins
$B =$ bandwidth in hertz
Therefore noise power due to R_G is

$$\frac{\bar{e}_T^2}{R_G} = \frac{4kTR_GB}{R_G} = 4kTB \tag{3-95}$$

The noise power of the FET referred to the input is

$$\frac{\bar{e}_n^2}{R_G} + \bar{i}_n^2 \cdot R_G \tag{3-96}$$

When expressions for the noise power of both the FET and R_G are substituted, the noise factor becomes

$$F = 1 + \frac{\bar{e}_n + \bar{i}_n^2 R_G^2}{4kTR_GB} \tag{3-97}$$

A noise figure NF expressed in decibels indicates the presence of added noise power from the FET or another active device. The noise figure is always given with reference to a standard, specifically the generator resistance R_G:

$$NF = 10 \log F \qquad (3\text{-}98)$$

The noise figure of the FET is

$$NF = 10 \log \left(1 + \frac{\bar{e}_n^2 + \bar{i}_n^2 R_G^2}{4kTR_G B} \right) \qquad dB \qquad (3\text{-}99)$$

When junction FET noise is expressed in terms of the noise figure NF, an inherent disadvantage arises in that the noise figure is dependent upon the value of the generator resistance R_G. Therefore, the $\bar{e}_n$, $\bar{i}_n$ method remains as the best way to quantitatively express the noise characteristics of the FET itself.

3-12-2   Describing Junction FET Noise Characteristics

Junction FET $\bar{e}_n$ and $\bar{i}_n$ characteristics are frequency-dependent within the audio noise spectrum, and take a form as shown in Fig. 3-54. $\bar{e}_n$, the equivalent short-circuit input noise voltage (with the exception of the $1/f^n$ region), is defined as

$$\bar{e}_n = \sqrt{4kTR_N B} \qquad (3\text{-}100)$$

where $R_N \cong 0.67/g_{fs}$, the equivalent resistance for noise. The $\bar{e}_n$, except

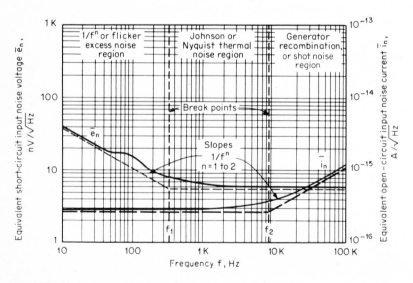

FIGURE 3-54   Characteristics of junction FET noise.

in the $1/f^n$ region, closely approximates the equivalent thermal noise voltage of the channel resistance.

In the so-called $1/f^n$ region, $\bar{e}_n$ is expressed as

$$\bar{e}_n = \sqrt{4KR_NB\left(1 + \frac{f_1}{f^n}\right)} \qquad (3\text{-}101)$$

where n varies between 1 and 2 and is device- and lot-oriented.

The characteristic bulge in $\bar{e}_n$ in the $1/f^n$ region has been observed to some extent in all junction FETs submitted to test. The breakpoint or corner frequency shown as f_1 in Fig. 3-54 is lot- and device-design–oriented, and varies from about 100 Hz to 1 kHz.

As indicated in Eqs. (3-100) and (3-101), $\bar{e}_n$ is inversely proportional to the square root of the transconductance of the FET ($e_n \propto 1/\sqrt{g_{fs}}$). $\bar{e}_n$ can be lowered by a factor of $1/\sqrt{N}$ if N devices with matched electrical characteristics are connected parallel. For example, when

$$N = 2 \qquad (3\text{-}102)$$

let

$$\bar{e}_{n1} = \bar{e}_{n2} \qquad (3\text{-}103)$$

and let

$$g_{fs1} = g_{fs2} \qquad (3\text{-}104)$$

Thus,

$$g_{fs(total)} = 2g_{fs1} \text{ or } 2g_{fs2} \qquad (3\text{-}105)$$

From Eq. (3-100)

$$\bar{e}_{n1} \doteq \sqrt{4kT\left(\frac{0.67}{2g_{fs1}}\right)B} \qquad (3\text{-}106)$$

and

$$\bar{e}_{n(total)} = \sqrt{4kT\left(\frac{0.67}{2g_{fs1}}\right)B} \qquad (3\text{-}107)$$

Thus,

$$\bar{e}_{n(total)} = \frac{1}{\sqrt{2}}\,\bar{e}_{n1} \qquad (3\text{-}108)$$

A second way to achieve low $\bar{e}_n$ is to use a device with a large gate area. Empirically, $\bar{e}_n$ is inversely proportional to the square of the gate area ($e_n \propto 1/A_G^2$), independent of g_{fs}. This large gate area philosophy has been followed in the design of the Siliconix 2N4867A FET, and noise performance of the device is discussed later. A major advantage of this type of design is that $\bar{e}_n$ is significantly lowered. $\bar{i}_n$ also remains at a low value.

The equivalent open-circuit input noise current $\bar{i}_n$, except in the shot noise region shown in Fig. 3-54, is due to thermally generated reverse current in the gate-channel junction. It is defined as

$$\bar{i}_n = \sqrt{2qI_GB} \qquad (3\text{-}109)$$

where $q = 1.602 \times 10^{-19}$ C (the magnitude of the electron charge)
I_G = measured dc operating gate current in amperes
B = bandwidth in hertz

The expression is accurate only when the measured gate current is the result of bulk device conductance. It is possible for the measured gate current to be due to conductance stemming from contamination across the leads of the semiconductor package.

At higher frequencies, as in the shot-noise region shown in Fig. 3-54, $\bar{i}_n$ can be approximated as being equal to the Nyquist thermal noise current generated by a resistor:

$$\bar{i}_n = \sqrt{\frac{4kTB}{R_p}} \qquad (3\text{-}110)$$

where R_p is the real part of the gate-to-source input impedance. The breakpoint of corner frequency f_2 in Fig. 3-54 is lot- and device-design–oriented. It usually is between 5 and 50 kHz.

Another form of noise found in junction FETs is known as "popcorn" or burst noise; the term popcorn noise was originated in the hearing aid industry because of noise or level shifts which are present in input stages and which resemble the sound of corn popping. Popcorn noise is a form of random burst input noise current which remains at the same amplitude and which is confined to frequencies of 10 Hz or lower. The suitability of a FET device is dependent on the amplitude of the burst, its duration, and its repetition rate. The origins of popcorn noise are not completely identified, but it is believed to be caused by intermittent contact in aluminum-silicon interfaces and by contamination in the oxidation process.

A test circuit to measure popcorn noise in differential junction FET amplifiers is shown in Fig. 3-55. In practice, popcorn noise is evaluated on an engineering basis, not on a production-line basis. There is no apparent correlation between $1/f^n$ noise at 10 Hz and popcorn noise. However, if the amplitude of the burst is large and occurs frequently, then $1/f^n$ noise voltage $(\bar{e}_n)$ is masked and difficult to evaluate at 10 Hz. The graph in Fig. 3-56 shows "moderate" burst noise observed in

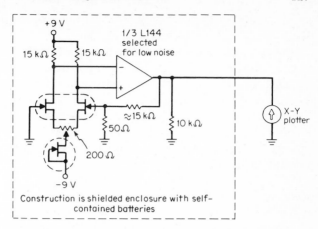

FIGURE 3-55   Test circuit to measure popcorn noise.

a group of junction FET differential amplifiers which were measured in the test circuit.

3-12-3   Operating-Point Considerations

Unlike bipolar transistors, where $\bar{e}_n$ and $\bar{i}_n$ characteristics vary directly with change in collector current I_C, similar characteristics in junction FETs will vary only slightly as drain current I_D is varied. This is true

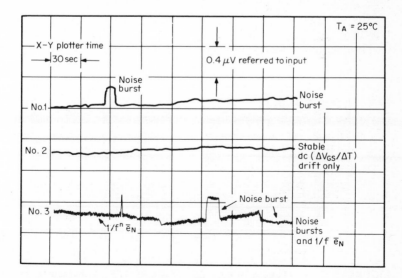

FIGURE 3-56   Popcorn noise in differential amplifiers.

so long as the FET is biased so that the drain-source gate is greater than the pinchoff voltage ($V_{DG} > V_{GS(off)}$).

The e_n in junction FETs will be lowest when the devices are operated at $V_{GS} = 0(I_D = I_{DSS})$, where transconductance g_{fs} is at its highest value. This will be true only if device dissipation is maintained very low in relation to the total dissipation capability of the FET.

The curves in Fig. 3-57 illustrate changes in $\bar{e}_n$ as the operating drain current I_D is varied. Note that the lowest $\bar{e}_n$ did not occur at $V_{GS} = 0$, because of high power dissipation and a resultant rise in junction temperature at the operating point.

The optimum (lowest) $\bar{i}_n$ in depletion-mode junction FETs should occur at $V_{GS} = 0(I_D = I_{DSS})$. In practice, very little change will be seen in $\bar{i}_n$ when the operating point is changed, provided that the drain-gate voltage is maintained below the I_G breakpoint and power dissipation is kept at a low level. The curves in Fig. 3-58 illustrate $\bar{i}_n$ characteristics as a function of drain-gate voltage at points below, on, and above the I_G breakpoint voltage.

To minimize I_G under operating conditions, particular attention must be paid to V_{DG}. The critical drain-gate voltage (I_G breakpoint voltage) can be anywhere from 8 to 40 V, depending on device design.

Gate operating current I_G should not be considered equal to gate I_{GSS} in linear amplifier applications. I_{GSS} is the reverse-biased gate-junction leakage current with $V_{DS} = 0$. The curves in Fig. 3-59 show how I_G breakpoint is related to basic device design. Device designs with a high g_{fs}/C_{iss} ratio typically have low breakpoint voltages, at $V_{DG} = 10$ V, whereas high-μ devices ($\mu = r_{ds} \cdot g_{fs}$) have much higher I_G breakpoints, typically $V_{DG} = 20$ to 30 V.

Three equations presented earlier show that $\bar{e}_n$ and $\bar{i}_n$ are proportional

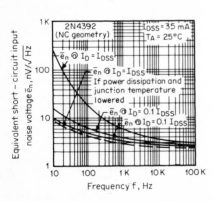

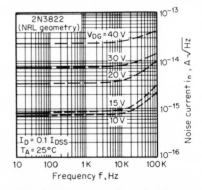

FIGURE 3-57 $\bar{e}_n$ changes versus I_D variations.

FIGURE 3-58 $\bar{i}_n$ characteristics as a function of drain-gate voltage.

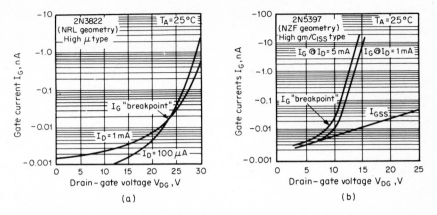

FIGURE 3-59   Gate operating current vs. drain-gate voltage.

to $\sqrt{T}$. Both will be reduced if the temperature is lowered. In Eq. (3-109), $\bar{i}_n$ is proportional to $\sqrt{I_G}$. If I_G is below its breakpoint, I_G will halve for each temperature drop of 10 to 11°C. $\bar{e}_n$ is also proportional to $\sqrt{R_N}$, where $R_N \cong 0.67/g_{fs}$. Thus when g_{fs} is increased, which is typical of junction FETs operating at low temperature, $\bar{e}_n$ will also be reduced.

Fig. 3-60 shows g_{fs} versus temperature for a silicon junction FET. Performance improves with decreasing temperature until about 100 K is reached. Below 100 K g_{fs} drops rapidly.

In connection with the plot of g_{fs} versus temperature, note that the relationship can vary from approximately 0.2 to 1 percent per degree Celsius. The g_{fs} slope depends upon the basic design of the FET and upon the proximity of the drain-current operating point to I_{DZ}, the zero temperature coefficient point.

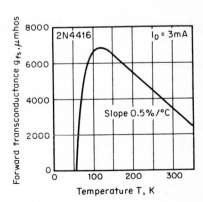

FIGURE 3-60 g_{fs} versus temperature.

The major application for junction FETs at low temperature is in charge-sensitive amplifiers. For best performance in this type of application, a high g_{fs}/C_{iss} ratio is required. Recommended Siliconix FET types for such applications are the 2N4416 (NH geometry) and the U311 (NZA geometry).

3-12-4 Test Measurements

By definition, $\bar{e}_n$ and $\bar{i}_n$ are referred to the input of the device under test. To measure $\bar{e}_n$, the test circuit shown in Fig. 3-61 is useful.

The following procedure should be used to make the $\bar{e}_n$ test:

1. Set tunable filter to required f_{low} and f_{high}. Adjust oscillator to mean center frequency

$$[f_{mean} = (f_{low} \cdot f_{high})^{1/2}]$$

2. Set V_{osc} to 100 mV with switch 1 in position 1. Compute

$$v_{in1} = 10^{-1} \times \frac{10^2}{16^6} = 10^{-5}\,V = 10\,\mu V$$

3. Measure V_{out1}. Compute overall gain as

$$A_V = \frac{V_{out1}}{V_{in1}} = \frac{V_{out1}}{10\,\mu V}$$

4. Set switch 1 to position 2 and measure V_{out2}. Compute V_{in2}, the equivalent short-circuit input noise voltage $\bar{e}_n$, using A_V from Step 3.

$$V_{in2} = \frac{V_{out2}}{A_V} = \bar{e}_n$$

in volts over bandwidth f_{low} to f_{high}.

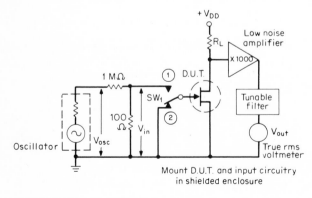

FIGURE 3-61 Test circuit to measure $\bar{e}_n$.

An alternative method of performing the above test is to use a Quan-Tech Transistor Noise Analyzer consisting of a Model 2173 Control Unit and a Model 2181 Filter. The analyzer has provision for measuring $\bar{e}_n$ and determining NF with various values of R_G in FET and bipolar devices with selectable test conditions. The measuring system has a constant gain of 10,000. The analyzer records output noise at selected frequencies between 10 and 100 Hz in the device under test, with the scale shown as the actual output divided by 10,000. This is then the output noise referred to the input. The equivalent bandwidth for testing is 1 Hz.

There are certain instances where the test circuit or the Transistor Noise Analyzer is not adequate to measure $\bar{e}_n$ at certain frequencies over certain bandwidths in the $1/f^n$ region. The rms noise over a bandwidth from f_{low} to f_{high}, where there is a $1/f^n$ characteristic over the entire range, can be computed as

$$\bar{e}_n = (e_n \text{ known}) \left[f_{known} \cdot \ln\left(\frac{f_{high}}{f_{low}}\right) \right]^{1/2n} \tag{3-111}$$

Figure 3-62 represents this equation graphically. For example, $\bar{e}_n$ known $= 70 \times 10^{-9}$ V/$\sqrt{\text{Hz}}$ at 10 Hz. How much noise is in the band from 4.5 to 5.5 Hz? The noise has a $1/f^1$ characteristic over the entire range. Thus

$$\bar{e}_n = (70 \times 10^{-9}) \left[10 \cdot \ln\left(\frac{5.5}{4.5}\right) \right]^{1/2} \text{ V} \tag{3-112}$$

or

$$\bar{e}_n = 99.16 \times 10^{-9} \text{ V}/\sqrt{\text{Hz}} \text{ @ } 4.975 \text{ Hz} \tag{3-113}$$

4.975 Hz is the mean center frequency where $f_{mean} = (f_{low} \cdot f_{high})^{1/2}$.

$\bar{i}_n$ measurements are difficult to implement at best. At frequencies below f_2 in Fig. 3-55, i_n is assumed to have a constant level or "white"

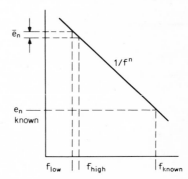

FIGURE 3-62 Computing rms noise over a bandwidth.

noise characteristic which may be correlated with gate current I_G. From Eq. (3-109) I_G is established as the measured bulk gate current. Because measured gate current I_G is the result of all conductances at the gate, the resultant gate current and the computed $\bar{i}_n$ due to bulk material can be assumed to be this value or less. The total equivalent input noise of the FET can be approximated by

$$\bar{e}_{ni}^2 = \bar{e}_T^2 + \bar{e}_n^2 + \bar{i}_n^2 \, R_G^2 \qquad (3\text{-}114)$$

$\bar{e}_T$ = thermal noise of the generator resistance R_G

$\bar{e}_{ni}$ = total FET noise referred to the input

This approximation assumes that the equivalent noise voltage and the current generators vary independently. Equation (3-114) implies that $\bar{i}_n$ can be calculated if $\bar{e}_n$, $\bar{e}_T$, and total noise $\bar{e}_{ni}$ are known. The difficulty here is that in MOS or junction FETs, the R_G must be very large to detect the anticipated small value of $\bar{i}_n$. However, when R_G is very large, $\bar{e}_T$ is much greater than $\bar{i}_n^2 \cdot R_G^2$. For example, over a 1-Hz bandwidth at 25°C, if R_G is equal to 100 MΩ, then

$$\bar{e}_T^2 = 4kTR_G = 4 \times 1.38 \times 10^{-23} \times 2.95 \times 10^2 \times 10^8$$
$$= 1.63 \times 10^{-12} \, \text{V}/\sqrt{\text{Hz}}$$

Anticipated $\bar{i}_n$ is

$$\bar{i}_n \simeq 10^{-15} \, \text{A}/\sqrt{\text{Hz}}$$

and

$$\bar{i}_n^2 = 10^{-30} \, \text{A}/\sqrt{\text{Hz}}$$

Thus

$$\bar{i}_n^2 \cdot R_G^2 = 10^{-30} \cdot 10^{16} = 10^{-14} \, \text{V}/\sqrt{\text{Hz}}$$

Therefore, $\bar{i}_n^2 \cdot R_G^2$ is much less than $\bar{e}_t^2$, which renders this method of finding $\bar{i}_n$ impractical for most common MOSFETs or junction FETs.

An improved method of measuring $\bar{i}_n^2$ is to substitute a low-loss mica capacitor for resistor R_G. The mica capacitor by definition does not have an equivalent thermal noise voltage, and thus Eq. (3-114) becomes

$$\bar{e}_{ni}^2 = \bar{e}_n^2 + \bar{i}_n^2 \cdot X_C^2 \qquad (3\text{-}115)$$

where X_C = capacitive reactance

or

$$\bar{i}_n = \frac{(\bar{e}_{ni}^2 - \bar{e}_n^2)^{1/2}}{X_C} \qquad (3\text{-}116)$$

When a 10-pF mica capacitor was used in the evaluation circuit (up to a frequency of 100 Hz), a correlation of from 80 to 90 percent was obtained when compared with $\bar{i}_n$ computed from measured gate current readings. At frequencies above 100 Hz direct computation of $\bar{i}_n$ via the capacitor method becomes unwieldy because of the rapid decrease in capacitor reactance at these frequencies.

In calculating $\bar{i}_n$ at higher frequencies, an alternative method is to measure R_p, the real part of the gate-source impedance of the FET. When R_p is measured at various frequencies, the equivalent short-circuit input noise current $\bar{i}_n$ can be computed as a function of frequency. [See Eq. (3-110).] A convenient instrument for measuring R_p is the Hewlett-Packard Type 250A Rx meter or equivalent. The Type 250A Rx meter can measure R_p accurately up to 200 kΩ. As is shown in Fig. 3-63, this establishes the low-frequency limit of 20 MHz for $\bar{i}_n$ computed via direct measurement of R_p for the Siliconix FET Type 2N4117A. For frequencies between 100 Hz and 20 MHz, $\bar{i}_n$ must be extrapolated, as is shown in Figs. 3-63 and 3-64. For FET types with lower R_p (such as the Siliconix 2N4393), $\bar{i}_n$ can be computed down to 2 MHz, and hence extrapolated $\bar{i}_n$ between 100 Hz and 100 kHz is more accurate.

Figure 3-65 shows representative $\bar{e}_n$, $\bar{i}_n$ curves for Siliconix JFET products. Of particular importance is the geometry which by its design governs the basic noise characteristics of product types derived from it.

3-12-5 Conclusion

Contemporary junction FETs have noise voltages $\bar{e}_n$ equal to those found in low-noise bipolar transistors. These two device types have different operating mechanisms: the FET is voltage-actuated, while the bipolar transistor is current-actuated. Hence, FETs have an inherently lower noise current $\bar{i}_n$ and are preferred over bipolar devices in most audio-frequency applications where low-noise performance is a design requirement.

When bias points are properly selected, as described in this chapter,

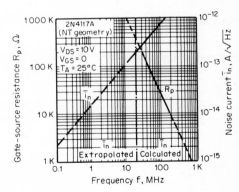

FIGURE 3-63 Low-frequency limit for calculated $\bar{i}_n$.

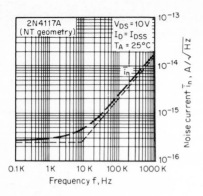

FIGURE 3-64 Extrapolated $\bar{i}_n$ versus frequency.

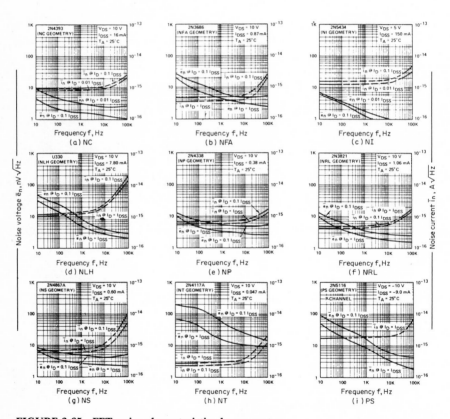

FIGURE 3-65 FET noise characteristics by geometry.

the excellent low-noise characteristics of high-g_{fs} junction FETs can be realized.

The curves shown in Fig. 3-65 are representatives of $\bar{e}_n$ and $\bar{i}_n$ performance of Siliconix junction FETs. Of particular importance in these curves is the process geometry by which the basic design of the FET governs the noise characteristics of product types derived from it. Readers are invited to refer to the Siliconix FET catalog for full geometry performance data and for specific part numbers stemming from the generic process geometries.

In the measurement section, it was shown that direct $\bar{e}_n$ measurements can readily be made. $\bar{i}_n$ can be guaranteed at frequencies below 100 Hz by measuring the dc operating gate current I_G. When I_G is known, $\bar{i}_n$ can be extrapolated from frequencies below 100 Hz to predict noise performance at frequencies to 100 kHz.

4

HIGH-FREQUENCY CIRCUITS

4-1   High-Frequency Techniques

4-2   The Problem of High-Frequency Amplification

4-3   The "Perfect" High-Frequency FET

4-4   Stability

4-5   Power Gain

4-6   Two-Port Transistor Parameters

4-7   Amplifiers

4-8   Mixers

4-9   Balanced Mixers

4-10   Double-Balanced Mixers

4-11   Oscillators

4-12   The High-Frequency Power FET

4-13   Noise Temperature and Noise Figure

4-1   HIGH-FREQUENCY TECHNIQUES

Field-effect transistors suitable for high-frequency applications have become aggressive competitors to bipolar transistors for both small-signal and power applications.

The first popular small-signal JFET suitable for VHF performance was the n-channel 2N3823 introduced by Texas Instruments in 1965. This device was characterized through 100 MHz. When Union Carbide (now Solitron) introduced their 2N4416 a year later, characterized through 400 MHz, it became the favorite for the remainder of the decade. It too had its faults, and although it maintained a lead position among high-frequency JFETs, it was not the panacea for trouble-free performance. It was close, but it missed the mark, and so there was room for the introduction of the MOSFET.

RCA introduced the first dual-gate MOSFET, the 3N140, in 1968, but not until their 3N187 came along two years later did the MOSFET become widely used in the TV and FM industry.

137

Both the n-channel JFET and the dual-gate MOSFET have become standard components in TV, FM, auto radios, guided missiles, radars, and wherever high-performance, high-frequency communications is found. Features and applications of some of the more prominent devices will be discussed further in this chapter.

In small-signal high-frequency applications the FET has several inherent advantages over the bipolar transistor, such as:

○ Square-law transfer characteristics which give at least a tenfold increase in input voltage handling capability with less spurious signal generation.

○ Good low-noise performance over the whole frequency range. This performance is especially noticeable in oscillators where low phase noise is of paramount importance.

○ Reduced sensitivity of input admittance to shifts in the operating point caused by AGC action, thus effecting less detuning of tuned circuits. The prime device for such applications is the dual-gate MOSFET, where 50 dB AGC can be achieved with insignificant detuning of high-Q tank circuits.

○ A high input resistance, in the common-source configuration, reduces the loading effects on input circuits, helping, among other things, to maintain high-selectivity tuned circuits.

○ Linear group delay over appreciable bandwidths and under varying AGC. Parasitic FET reactances are voltage-dependent and not generally affected by input signals.

For power FETs there are other advantages which are usually overlooked in small-signal FETs. The absence of minority-carrier storage time and of secondary breakdown is inherent in both JFETs and MOSFETs. For most bias conditions the drain current has a negative temperature coefficient; thus thermal runaway is not a likely problem. This latter characteristic simplifies the paralleling of power FETs to increase drain-current capabilities; "current hogging" does not occur. Their high-speed switching capabilities permit increased operating efficiencies in high-frequency switch-mode operation (classes D, E, and F). Their high-frequency characteristics also offer freedom from load mismatches (infinite VSWR* capabilities).

FETs made with gallium arsenide have made sizable encroachments

---

* VSWR = voltage standing wave ratio.

into the microwave bipolar transistor market. Recently developed power FETs such as the vertical-channel MOS (VMOS) will replace power bipolar transistors in many applications.

4-2 THE PROBLEM OF HIGH-FREQUENCY AMPLIFICATION

It was not until the manufacturers had refined their processes to achieve FETs with high transconductances and low interelectrode and parasitic capacitances that there emerged a suitable high-frequency device.

As the operating frequency is increased, the effects of parasitic capacitances become the fundamental restriction upon the FET—in particular, the effect on its input impedance. In the common-source stage, the Miller effect becomes the dominant capacitance. Input impedance is largely a reactance of C_{in}:

$$X_{in} = \frac{1}{2\pi f C_{in}} \tag{4-1}$$

where $C_{in} = C_{iss} - A_V C_{rss}$

For the common-source amplifier the quantity A_V is negative (due to the phase reversal); therefore

$$
\begin{aligned}
C_{in} &= C_{iss} + |A_V C_{rss}| \\
&= C_{gs} + C_{gd} + |A_V C_{dg}|
\end{aligned}
\tag{4-2}
$$

The reverse capacitance C_{rss} of a FET, in addition to raising the input capacitance through the Miller effect, feeds back signals from the output to the input, contributing to possible instability—especially if the load itself is reactive.

Much of the design effort in high-frequency amplifiers is directed toward reduction of the effects of the feedback capacitance. Although the common-gate amplifier has lower power gain than the common-source type, it is widely used at high frequencies because of its much lower feedback capacitance. The magnitude of the voltage gain for these two types of amplifiers may be comparable; however, in the common-source stage, C_{gd} is the feedback capacitor, while in the common-gate stage it is C_{ds}. Also A_V for the common gate is positive; thus, for the common gate,

$$C_{in} = C_{gs} + C_{ds} - |A_V| C_{ds} \tag{4-3}$$

For many device types C_{ds} is an order of magnitude lower than C_{gd}. The resulting reduction in the feedback capacitance makes the common-gate amplifier attractive despite its lower power gain. Other advantages will be examined in this chapter.

A common figure of merit used to compare high-frequency performance of FETs is

$$\text{Figure of merit (FM)} = \frac{g_m}{2\pi C_{\text{in}}} \tag{4-4}$$

and for the popular n-channel JFET, the U310, this figure equates to 600 MHz. To illustrate the improvement in n-channel JFETs over the period of 1965 to 1976, the following should be of interest:

Type	FM relative to 2N4416
2N3823	0.6
2N4416	1.0
2N5397	1.3
U310	2.1

There are expressions of high-frequency performance other than the classic figure of merit [Eq. (4-4)] that should be considered. Although perhaps not immediately apparent, a "quality factor" expression

$$\text{QF} = \frac{g_{fs}}{g_{is}} \tag{4-5}$$

becomes especially important in the selection of JFETs suitable for ultra-broadband transmission-line (distribution) amplifiers.

DESIGN PRIORITIES FOR HIGH-FREQUENCY AMPLIFIERS

Amplification of	Analog signals	Digital signals
Small signal	High intercept	Flat group delay
	Low crossmods	High intercept
	Low noise figure	Low crossmods
	Sufficient gain	Low noise figure
	AGC	Sufficient gain
		AGC
Large signal	Linear	
	Low spurious	
	Low sideband noise	Same
	Overload protected	
	Mismatched loading	

In small-signal amplifiers an equally important ratio is that of forward transconductance to output conductance, more properly identified as mu.

$$mu = \frac{g_{fs}}{g_{os}} \qquad (4\text{-}6)$$

Each of these quality factors may serve as a guide in the selection of the proper transistor for a given application. Seldom does one achieve the perfect FET. What a "perfect" FET might be will be reviewed later in this chapter. Design priorities for high-frequency amplifiers can be outlined according to applications (see table on page 140).

FETs more closely meet these design objectives than do any other known active semiconductor. This chapter will continue to examine these priorities as they are applied to amplifiers, mixers, and oscillators.

4-3 THE "PERFECT" HIGH-FREQUENCY FET

There is no "perfect" high-frequency FET, nor is there likely to be one. Factors that inhibit optimum high-frequency gain and noise figure are: (1) feedback capacitance C_{gd}; (2) shunting input conductance g_{is}; and (3) output conductance g_{os}. High forward transconductance g_{fs} is of paramount importance, and its magnitude is directly affected by the magnitude of both the input and output conductances.

In the common-source amplifier the input conductance g_{is} poses a mixed problem. Although a very low input conductance raises the quality factor (g_{fs}/g_{is}), which is desirable, a more debilitating effect occurs. A very low conductance (a high input resistance) infers high input Q, and effective coupling from the input circuitry may well result in excessive losses and increased overall noise figure in wideband circuits.

Output conductance g_{os}, on the other hand, directly affects the forward gain of the FET. Maximum available voltage gain is fundamentally related to the ratio

$$\frac{g_{fs}}{g_{os}}$$

An appreciation of how g_{fs} must be improved to overcome the debilitating effects of both the shunting effects of increasing g_{is} and the reduced mu caused by the increasing g_{os} is provided in Fig. 4-1.

The chart in Fig. 4-1 is interpreted as follows: P_{GU} is unilateralized power gain expressed in decibels. But note carefully that each value is underlined. For example, 10 dB (solid underline) and 15 dB (dashed underline).

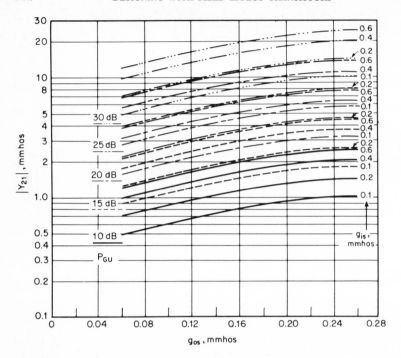

FIGURE 4-1 Projected UHF FET transconductance as a function of g_{is}, g_{os}, and gain. Unilateralized power gain common source ($Y_{12} = 0$).

The value of g_{is} (right vertical axis) ranges from 0.1 to 0.6 mmho and is repeated five times. Again, note that each time a value of g_{is} is repeated, it is to identify a different coded graph that corresponds to the underlined value of P_{GU}.

For example, assume we desire a unilateralized power gain P_{GU} of 20 dB. We have a value of g_{is} of 0.4 mmho and a value of g_{os} of 0.12 mmho. From this chart we determine that we need a value of $|Y_{21}|$ of 4.3 mmhos.

Feedback affects circuit stability, complicates circuit design, and reduces performance. Manufacturers reduce C_{gd} by careful attention to the metalizations and diffusions in the design of the FET chip as well as in chip packaging.

4-4 STABILITY

A very important consideration in the design of a small-signal, high-frequency amplifier is its stability, its freedom from a tendency to oscillate. Amplifier "stability factors" have been derived by Linvill[1] and by

Stern.[1] Linvill assumes that the amplifier stage has infinite source and load impedance, which are the worst-case stability conditions. Stern takes into account not only the transistor, but also the source and load impedances.

The Linvill stability factor C is expressed as

$$C = \frac{|Y_f Y_r|}{2 \operatorname{Re} Y_1 \operatorname{Re} Y_0 - \operatorname{Re}(Y_f Y_r)} \tag{4-7}$$

If C is less than 1, the transistor is considered to be unconditionally stable. If C is greater than 1, oscillation is possible if the transistor is presented with some specific source and load impedance; it is considered to be only conditionally stable. Unconditional stability assures that any positive real impedance presented to the transistor will not cause it to oscillate. This assumes no external feedback.

The Stern stability factor K is

$$K = \frac{2(g_{11} + g_s)(g_{22} + g_1)}{|Y_{12} Y_{21}| + \operatorname{Re}(Y_{12} Y_{21})} \tag{4-8}$$

In this expression it is assumed that the input and output matching networks are simultaneously tuned to the same frequency. The Stern solution defines $K > 1$ as being unconditionally stable. If g_s and g_1 (source and load conductance) are set equal to zero, the result is the reciprocal of the Linvill stability factor!

There are many ways to ensure stability in an amplifier. From the equations for the Linvill stability factor the following equations can be derived:

$$g_{\text{input loading}} = \frac{|Y_{21} Y_{12}|/C + \operatorname{Re}(Y_{21} Y_{12})}{2 g_{22}} - g_{11} \tag{4-9}$$

$$g_{\text{output loading}} = \frac{|Y_{21} Y_{12}|/C + \operatorname{Re}(Y_{21} Y_{12})}{2 g_{11}} - g_{22} \tag{4-10}$$

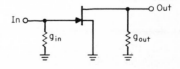

FIGURE 4-2 Two-port with conductance loading.

$g_{\text{input loading}}$ and/or $g_{\text{output loading}}$ are those values of conductance which must be added to the transistor externally, as in Fig. 4-2, to obtain the desired stability factor C.

In many cases it is not desirable to use these techniques to stabilize a transistor. Some of the power which would ordinarily be delivered to the load is now dissipated in the stabilizing resistors.

Stern developed a technique in which the input and output could be mismatched by equal amounts to assure stability of the amplifier. This technique causes a mismatched input and output as the price for circuit stability. In this case one establishes the Stern stability factor and then calculates by an iterative process the values for b_{in} and b_{out}.

$$G_s = \sqrt{\frac{K[|Y_{21}\,Y_{12}| + \mathrm{Re}(Y_{21}\,Y_{12})]}{2}} \; \sqrt{\frac{g_{11}}{g_{22}}} - g_{11} \qquad (4\text{-}11)$$

$$G_l = \sqrt{\frac{K[|Y_{21}\,Y_{12}| + \mathrm{Re}(Y_{21}\,Y_{12})]}{2}} \; \sqrt{\frac{g_{22}}{g_{11}}} - g_{22} \qquad (4\text{-}12)$$

where one can solve for both b_{in} and b_{out} (the *susceptance* values of Y_{in} and Y_{out}) by

$$Y_{in} = Y_{11} - \frac{Y_{21}\,Y_{12}}{Y_{22} + Y_l} \qquad (4\text{-}13)$$

and

$$Y_{out} = Y_{22} - \frac{Y_{21}\,Y_{12}}{Y_{11} + Y_s} \qquad (4\text{-}14)$$

A third method of ensuring stability is by neutralization or unilateralization. Neutralization is a technique used to set b_{12} or the reactive component of the feedback y_{12} equal to zero. In most instances this would be equivalent to adding, between the input and output of the transistor, an inductor whose reactance is equal in magnitude to the reactance of the transistor feedback capacitance (C_{rss}). This is equivalent to placing a parallel tuned trap between the input and output matching networks.

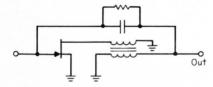

FIGURE 4-3 A two-port with unilateri-zation feedback.

Unilateralization is that technique in which g_{12} and b_{12} are set to zero. This might be implemented as shown in Fig. 4.3, by using a $1:1$ phase-inverting transformer at the input or output with a resistor and capacitor equal to g_{12} and b_{12} tied back across the transistor.

The advantages of neutralization and unilateralization are that the input and output are effectively isolated from each other and maximum stable gain results. The disadvantage is that in most instances (among small-signal amplifiers) the operating Q is increased, forcing a decrease in bandwidth.

Most JFETs and single-gate MOSFETs, when operating in their common-source configuration, are only conditionally stable ($C > 1$). Because they are conditionally stable, the precautions for stability must be observed. Alternatives to the common-source configuration are the common-gate and the Cascode configurations. When a JFET is operated in the common-source configuration, there is a relatively large intrinsic capacitance C_{gd} between input and output. However, when the JFET is used common-gate, C_{gd} is shunted to ground, and capacitance between input and output is the relatively small drain-to-source capacitance, generally in femtofarads.

An effective Cascode arrangement is the popular dual-gate MOSFET, where the isolation between input and output can reach 50 dB, thus affording considerable stability.

4-5 POWER GAIN

Power gain may be defined as the ratio of power delivered to the load to power absorbed at the input of the amplifier. Therefore, the power gain of a JFET common-source amplifier should be very high, as the input resistance is high, thus absorbing little power. Conversely, one can expect the common-gate amplifier to have reduced gain compared with its common-source counterpart. However, in the common-source configuration, because of stability requirements, the gain of the common-source JFET amplifier must be reduced.

There are many things to consider when designing an amplifier for a specific power gain. More often than not, system gain requirements may be less than the gain capability of the transistor. For example, a required stage gain of 10 dB is needed from a JFET that could easily offer 15 dB.

Given the expression for transducer gain[1]

$$G_t = \frac{4g_{11}g_{22}\,|Y_{21}|^2}{|(Y_{11} + Y_s)(Y_{22} + Y_1) - Y_{12}Y_{21}|^2} \tag{4-15}$$

the power gain of an amplifier can be determined. G_t (transducer power gain) is the gain of an amplifier when the input and output matching networks, as well as the transistor, are considered. In general, gain can be reduced by lowering the transconductance of the FET, or the load impedance or the source impedance.

FET gain is generally attributed to its forward transconductance. As mentioned in earlier chapters, a FET has two fundamental operating areas: the triode area, where $r_{DS(on)}$ predominates, and the pentode area of current saturation. Controlling transconductance by the manipulation of current is tedious and not recommended. When the need to lower the gain arises, it is often beneficial to manipulate the input impedance to bring it closer to the source impedance and thus provide a better match.

If the transistor is unconditionally stable, maximum available gain (MAG) is

$$\text{MAG} = \frac{|Y_{21}|^2}{A + (A^2 - |Y_{12}Y_{21}|^2)^{1/2}} \tag{4-16}$$

where $A = 2g_{11}g_{22} - \text{Re}(Y_{21}Y_{12})$

To obtain MAG both the input and output must be a conjugate match. For these conditions,

$$g_s = \frac{1}{2g_{22}}(A^2 - |Y_{21}Y_{12}|^2)^{1/2} \tag{4-17}$$

$$b_s = -b_{11} + \frac{\text{Im}(Y_{21}Y_{12})}{2g_{22}} \tag{4-18}$$

$$g_1 = \frac{1}{2g_{11}}(A^2 - |Y_{21}Y_{12}|^2)^{1/2} \tag{4-19}$$

and

$$b_1 = -b_{22} + \frac{\text{Im}(Y_{21}Y_{12})}{2g_{11}} \tag{4-20}$$

where $A = 2g_{11}g_{22} - \text{Re}(Y_{21}Y_{12})$

There are many situations where G_{max} is greater than it is required to be. In such cases, Stern offers a technique by which the input and output are equally mismatched to obtain a desired stability factor and a gain less than MAG. Linvill has described a technique where only the output is mismatched and the input is conjugately matched. Both Stern and Linvill lend themselves to graphical solutions.

4-6   TWO-PORT TRANSISTOR PARAMETERS

4-6-1   Z, Y, H, and S Parameters

Transistors are characterized at high frequency using one of four possible parameter sets. Z (impedance) parameters correspond to open-circuit constraints on the unexcited terminal pair. Y (admittance) parameters correspond to a short circuit at the unexcited terminal pair.

H (hybrid) parameters are a mixture of Z and Y parameters. The H parameter set is popular for bipolar transistor characterization because of the widely divergent impedances of bipolar transistors. In particular, for the common-base and common-emitter configurations, it is impractical to characterize the transistor with either Z or Y parameters. Consequently, H parameters have different dimensions.

S (scattering) parameters suggest that an incident signal impinging upon a mismatched load scatters into different components. If so, then S parameters provide a measure of this separation of scattered elements. Recently high-frequency design has focused heavily on S parameters rather than the formerly popular Y parameters. Both S and Y will be briefly discussed in this chapter. Z parameters and H parameters, which are more popular among bipolar transistor designers, will be omitted.

4-6-2   *Y* Parameters

Short-circuit admittances are still widely used in high-frequency calculations. The equations describing the common-source equivalent circuit of Fig. 4-4 are

$$I_{\text{in}} = Y_{is}E_{\text{in}} + Y_{rs}E_{\text{out}} \tag{4-21}$$

$$I_{\text{out}} = Y_{fs}E_{\text{in}} + Y_{os}E_{\text{out}} \tag{4-22}$$

where E_{in} and I_{in} are the input signal voltage and current and E_{out} and I_{out} are the output signal voltage and current.

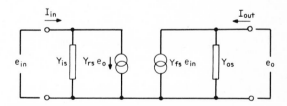

FIGURE 4-4   High-frequency admittance parameter JFET model.

Short-circuiting the output ($E_{\text{out}} = 0$) yields

$$Y_{is} = \frac{I_{\text{in}}}{E_{\text{in}}} \tag{4-23}$$

and

$$Y_{fs} = \frac{I_{\text{out}}}{E_{\text{in}}} \tag{4-24}$$

Short-circuiting the input ($E_{in} = 0$) yields

$$Y_{rs} = \frac{I_{in}}{E_{out}} \tag{4-25}$$

and

$$Y_{os} = \frac{I_{out}}{E_{out}} \tag{4-26}$$

thus

$$\begin{vmatrix} Y_{is} & Y_{rs} \\ Y_{fs} & Y_{os} \end{vmatrix} = Y_{is}Y_{os} \tag{4-27}$$

since

$$\frac{I_{in}}{E_{in}} \times \frac{I_{out}}{E_{out}} = \frac{I_{out}}{E_{in}} \times \frac{I_{in}}{E_{out}} \tag{4-28}$$

then

$$Y_{is}Y_{os} - Y_{rs}Y_{fs} = 0$$

Figure 4-5 shows a neutralizing path added to the FET. Adding this to the FET Y parameters, we arrive at

$$\begin{vmatrix} Y_{is} & Y_{rs} \\ Y_{fs} & Y_{os} \end{vmatrix} + \begin{vmatrix} Y_n & -Y_n \\ -Y_n & Y_n \end{vmatrix} = \begin{vmatrix} Y_{is} + Y_n & Y_{rs} - Y_n \\ Y_{fs} - Y_n & Y_{os} + Y_n \end{vmatrix} = 0$$

so

$$(Y_{is} + Y_n)(Y_{os} + Y_n) = (Y_{rs} - Y_n)(Y_{fs} - Y_n) \tag{4-29}$$

and

$$I_{in} = (Y_{is} + Y_n)E_{in} + (Y_{rs} - Y_n)E_{out} \tag{4-30}$$

$$I_{out} = (Y_{fs} - Y_n)E_{in} + (Y_{os} + Y_n)E_{out} \tag{4-31}$$

Since in neutralizing we equate $Y_{rs} = Y_n$, then

$$I_{in} = (Y_{is} + Y_{rs})E_{in} + 0 \tag{4-32}$$

$$I_{out} = (Y_{fs} - Y_{rs})E_{in} + (Y_{os} + Y_{rs})E_{out} \tag{4-33}$$

Since I_{in} is unaffected by E_{out}, neutralizing is effective in cancelling all feedback effects. In practice, Y_{rs} is largely capacitive, making the neutral-

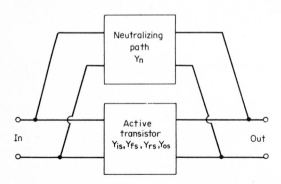

FIGURE 4-5   Basic two-port with feedback neutralization.

izing path largely inductive. Both paths—through the FET and through the neutralizing path—must have the same reactance at the frequency for which the amplifier is to be designed. That is,

$$2\pi fL = \frac{1}{2\pi fC_{gd}} \qquad (4\text{-}34)$$

or

$$L = \frac{1}{4\pi^2 f^2 C_{gd}} \qquad (4\text{-}35)$$

Each Y parameter is a complex quantity and may be derived from a "physical" equivalent circuit somewhat more comprehensive than that shown for low frequencies.

Y parameters are functions of frequency and are so described in most high-frequency data sheets. Typical of such graphs are those of Fig. 4-6, which relate to the Siliconix U310 JFET. Notice that each Y parameter is specified in terms of its real and imaginary components. Such data are most useful for computer-aided design (CAD).

Many FET data sheets show Y parameters extending to 1 GHz. However, in the conventional TO46, 52, or 72 package, performance above 600 MHz is limited because of the package capacitances and internal lead-bond inductances.

4-6-3 S Parameters

Scattering or S parameters are receiving increased attention in the characterization of devices and the design of high-frequency circuits. Modern computer-aided design (CAD) optimization programs, such as COMPACT,* rely upon the designer "inputting" S-parameter data. Some simplified S-parameter designs have omitted the important reverse transfer parameter S_{12} by equating it to zero ($S_{12} = 0$); although this greatly simplifies the "number crunching" of two-port network design, the results often have little utility.

When either the input scattering parameter S_{11} or the output scattering parameter S_{22} is displayed graphically, frequently superimposed on a Smith chart, one must be careful to remember that these scattering parameters are *reflection coefficients,* not impedance. Although both reflection coefficients and impedances can be plotted on a Smith chart, *they are by no means the same.*

Since the basic S parameter is derived from either voltage or current,

---

* COMPACT = Computerized Optimization of Microwave Passive and Active Circui Ts, Compact Engineering, Inc., Los Altos, CA 94022.

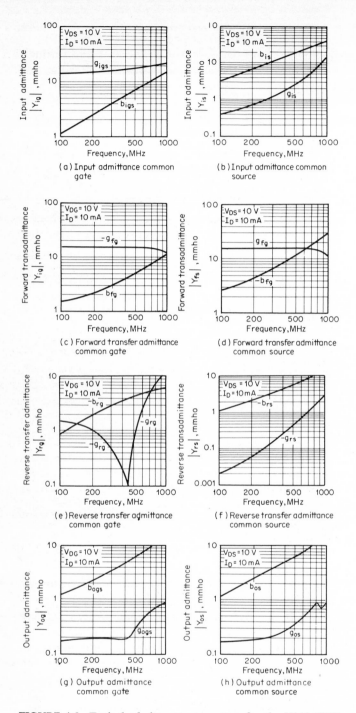

FIGURE 4-6 Typical admittance parameters for the U310 JFET.

it is correct to identify power parameters as the square of the S parameter. Consequently, for symmetrical networks the following identifications are noteworthy.

$$\% \text{ power reflected} = |S_{11}|^2 \times 100 \tag{4-36}$$

$$\% \text{ power transmitted} = |S_{21}|^2 \times 100 \tag{4-37}$$

$$\% \text{ power dissipated} = (1 - |S_{11}|^2 - |S_{21}|^2) \times 100 \tag{4-38}$$

A logical step shows that the forward power gain of a transistor is

$$G_{(dB)} = 10 \log |S_{21}|^2 \tag{4-39}$$

The forward operating power gain, taking into account reflections from the input port, is

$$G_{P(dB)} = 10 \log \left(\frac{|S_{21}|^2}{1 - |S_{11}|^2} \right) \tag{4-40}$$

Because of the increasing popularity and growing importance of S parameters, it is advisable for the reader to consult the massive literature on this subject.

4-7 AMPLIFIERS

The requirement for low cross modulation and intermodulation products makes the FET a practical necessity in solid-state RF circuitry. Recent improvements in the design of gallium arsenide FETs have extended their useful operating frequency into the gigahertz region, and the more conventional silicon JFETs to above 500 MHz. The Y parameters given in Fig. 4-6 are indicative of modern JFET performance at VHF/UHF.

The JFET can offer noise performance superior to many other forms of active amplifying elements, particularly at low frequencies. Of special interest is the effect of $1/f$ noise as related to RF amplifiers, which results in low phase jitter and low AM noise sidebands.

JFET amplifier design may use any of several possible configurations. The common source, the common gate, and the common drain (source follower) are the most universal. Each has its advantages and disadvantages. The Cascode design, being a combination of both the common source and common gate, offers distinct advantages. Each will be covered in this section.

4-7-1 The Common-Source Amplifier

In the common-source configuration (Fig. 4-7), the JFET exhibits several advantages, such as high input impedance, which, in turn, contributes

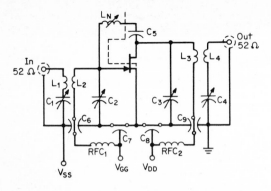

FIGURE 4-7 **450-MHz neutralized CS amplifier** (C_{1-4}—0.8- to 12-pF Johanson type 2950; C_5—40-pF DM5 silver mica; C_{6-9}—1000-pF Allen-Bradley type FA5C; L_1—1.4 in long, 22 gauge enamel, spaced 0.1 in from L_2; L_2—1.1 in long, 16 gauge solid copper; L_3—1.3 in long, 16 gauge solid copper; L_4—1.4 in long, 22 gauge enamel, spaced 0.3 in from L_3; $RFC_{1,2}$—0.15-μH Delevan type 1537-00; L_N—3 turns 22 gauge enamel, 0.25 in ceramic former, aluminum slug).

to high gain. A low noise figure is achieved with a noise-impedance match nearly equal to the power-gain impedance match. The major disadvantage of this common-source amplifier is that it is often potentially unstable, thus requiring special design precautions (see the section on stability).

An additional disadvantage of using the common-source configuration is its susceptibility to cross-modulation products which are voltage-dependent. Since the input impedance of the common-source amplifier is high, a signal of known input power will produce a high voltage on the gate, giving rise to potentially serious cross-modulation products.

The high input impedance of the common-source configuration allows a voltage step-up transformer to be used between the low-impedance line and the input to the JFET gate. Gain due to this voltage transformation may be expressed

$$A_V = \sqrt{\frac{R_{in}}{R_{gen}}} \tag{4-41}$$

If care is not taken, the low noise figure that may be expected in this circuit configuration will be lost due to the insertion loss inherent in the input network. A ratio of unloaded Q to loaded Q should be greater than $10:1$ to minimize this loss.

$$\text{Insertion loss (dB)} = -20 \log \left(1 - \frac{Q_l}{Q_u}\right) \qquad (4\text{-}42)$$

In Fig. 4-7 both the input and output circuits are capacitively tuned. At the input the interelectrode capacitance C_{iss} contributes significantly to the tuned network.

There are practical difficulties involved with neutralizing. Each circuit must be carefully adjusted for maximum gain at the desired frequency, and during the neutralizing procedure the signal fed back into the output port must be of a comparable magnitude, as will be anticipated in the output circuit during normal operation. Each amplifier must be individually neutralized for the specific JFET, as generally no two are alike. Shielding between input and output must be used for optimum performance. Quite often the success of the common-source neutralized amplifier hangs on the placement of this shielding!

4-7-2 The Common-Gate Amplifier

The common-gate amplifier is the most popular high-frequency JFET amplifier because:

- The low input resistance ($1/g_m$) together with the low input capacitance offers convenient matching to transmission lines and other low-impedance sources.

- The low input resistance offers excellent immunity to cross modulation.

- It is unconditionally stable, generally through UHF, due to low feedback.

Yet there are disadvantages; for example,

- Optimum noise impedance match does not offer optimum power gain.

- It cannot accept AGC without severe output-circuit detuning (not much different than any other JFET configuration).

- It has lower power gain than common source.

In many situations the JFET used in the common-gate (CG) configuration is unconditionally stable even at rather high frequencies (Fig. 4-8). However, care must be exercised in the design and layout of the CG amplifier; in particular, shielding between source and drain loads is critical. A

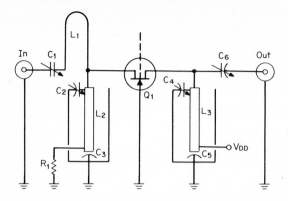

PARTS LIST FOR U310/NZA, 450-MHZ TEST FIXTURE

1.	Socket*	Augat	8060–1G9
2.	$L_2 L_3$**	NIBCO	Copper coupling 600, 1 in
3.	$C_1 C_2 C_4$	Johanson Manufacturing	0.8–10 pF, 5200
4.	$C_3 C_5$	Allen Bradley	1000-pF feedthrough
5.	C_6	Erie	0.25–1.5 pF, 530–000
6.	RF connector (two)	King	UG-1094
7.	L_1	—	14 awg bare copper, 2.8 in long with ⅜-in diameter bend, uniform lead length.
8.	Chassis	BUD	CU234

* Socket was divided to accept a copper foil shield soldered directly to both the gate lead and ground.
** Center conductor is ¼-in diameter copper tubing.

FIGURE 4-8   Test fixture U310/NZA—450 MHz.

JFET showing complete stability can become an oscillator if, for example, a low-loss socket is used without interelectrode shielding being maintained *within the socket itself!* A few femtofarads of coupling between socket pins can bring instability to an otherwise unconditionally stable amplifier.

The common-gate and common-source configurations have a common fault: their high drain impedance, which severely limits their performance over wide bandwidths. There are several solutions to achieve wide bandwidths, but none that does not limit the overall stage gain of the JFET amplifier. One easy solution is to load the drain circuit either by mismatch or by resistive loading; another means used in multistage amplifiers is stagger tuning; still another, but of little practical application in modern design, is the distributed amplifier design.

The common-gate amplifier, shown in Fig. 4-8, not having a frequency-dependent neutralization path, not only is free from the relevant adjustment problems, but is effective as a wideband amplifier. However, because of the lower input impedance, it does not offer the power gain of the common-source amplifier.

4-7-3 Common-Drain Amplifier (Source Follower)

The common-drain amplifier (sometimes called the source follower or voltage follower) characteristically has high input and low output impedances. It is frequently used to couple high-impedance video lines to low-impedance loads and is typically used at frequencies below 100 MHz.

The real part of the output impedance is $(g_{fs})^{-1}$, which is generally independent of frequency. If the load resistance is greater than $(g_{fs})^{-1}$, then C_{in} is also independent of frequency.

The high-frequency corner f_3 (where the response is down 3 dB) is an inverse function of the input time constant $R_g C_{in}$:

$$f_3 = \frac{1}{R_g C_{in}} \tag{4-43}$$

where

$$C_{in} = C_{gd} + C_{gs}(1 - A_V) + C_{stray} \tag{4-44}$$

and R_g is the input signal source resistance. Examples of how the signal source resistance affects bandwidth is shown in Fig. 4-9.

The voltage gain of the source follower is slightly less than unity and is given by

$$A_V = \frac{g_{fs} R_s}{1 + g_{fs} R_s} \tag{4-45}$$

From this we observe that the voltage gain is nearly independent of R_s when the $g_{fs} R_s$ product is large.

The output resistance of the FET, g_{fs}^{-1}, is in *parallel* with the source resistor R_S. Therefore, R_O of the amplifier is

$$R_O = \frac{R_S}{1 + g_{fs} R_S} \tag{4-46}$$

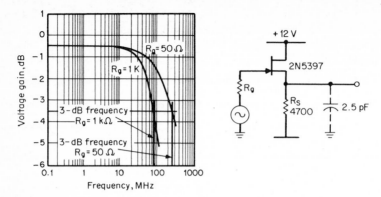

FIGURE 4-9   Effect of source impedance on bandwidth of a video amplifier.

4-7-4   The Cascode Amplifier

The FET Cascode circuit shown in Fig. 4-10 brings to solid-state amplifiers the high-frequency characteristics of the pentode vacuum tube, that is, reduced Miller effect resulting in improved stability and bandwidth. Although JFETs may be used in Cascode circuits, they have pretty much given way to the more popular dual-gate MOSFET, which is a monolithic Cascode element.

The Cascode amplifier, whether it is constructed using a pair of JFETs or a single dual-gate MOSFET, provides a tenfold reduction in feedback capacitance compared with a simple single-stage common-source JFET amplifier.

Practical advantages of this reduced Miller effect can be demonstrated in the gain-bandwidth calculations for an RC-coupled video amplifier. The gain-bandwidth product (GBW) is defined by

$$\text{GBW} = \frac{g_{fs}}{2\pi(C_{in} + C_{out})} \tag{4-47}$$

For the single common-source JFET, C_{in} may be calculated:

$$C_{in} = C_{gs} + C_{dg}(1 - A_V) \tag{4-48}$$

and

$$C_{out} = C_{dg} \tag{4-49}$$

Here the Miller-effect capacitance, $(1 - A_V)C_{dg}$, places a shunting input capacitance proportional to the stage gain. This limits high-frequency performance. In the Cascode amplifier stage the effective loading of the second stage upon the drain connection of the first stage limits the first stage gain to near unity (-1) and thereby forces the input capacitance C_{in} to be

$$C_{in} = C_{gs} + 2C_{dg} \qquad (4\text{-}50)$$

C_{iss} is $C_{gs} + C_{dg}$; consequently, $C_{in} = C_{iss} + C_{dg}$. The improvement of the gain-bandwidth product between a JFET and a dual-gate MOSFET can be most effectively shown by comparing the Siliconix U311 with the 3N201, both exhibiting near-equal forward transconductance g_{fs}.

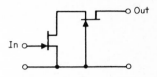

FIGURE 4-10 Basic Cascode configuration.

Operating the U311 in the common-source configuration, the typical parameters are:

$$g_{fs} = 14 \text{ mmhos}$$
$$C_{gs} = 3.2 \text{ pF}$$
$$C_{gd} = 1.2 \text{ pF}$$

The 3N201 parameters are:

$$g_{fs} = 14 \text{ mmhos}$$
$$C_{gs} = 4 \text{ pF}$$
$$C_{gd} = 0.2 \text{ pF (including strays)}$$

Beginning with the U311 with a load resistance of 2000 Ω,

$$A_V = g_{fs}R_1 = 14 \times 10^{-3} \times 2 \times 10^3 = 28$$
$$C_{in} = C_{gs} + C_{gd}(1 + A_V) = 3.2 \times 10^{-12} + (1.2 \times 10^{-12})(1 + 28) = 38 \text{ pF}$$
$$C_{out} = C_{gd} = 1.2 \times 10^{-12} \text{ F}$$

From Eq. (4-47),

$$\text{GBW} = \frac{14 \times 10^{-3}}{6.28(38 \times 10^{-12} + 1.2 \times 10^{-12})} = 56.8 \text{ MHz}$$

For the dual-gate MOSFET 3N201,

$$A_V = g_{fs}R_1 = 28$$
$$C_{in} = C_{gs} + 2C_{gd} = 4 \times 10^{-12} + 2(0.2 \times 10^{-12}) = 4.4 \text{ pF}$$
$$C_{out} = C_{gd} = 0.2 \times 10^{-12}$$
$$\text{GBW} = \frac{14 \times 10^{-3}}{6.28(4.4 \times 10^{-12} + 0.2 \times 10^{-12})} = 485 \text{ MHz}$$

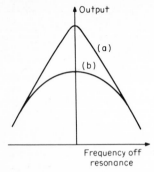

FIGURE 4-11 Amplifier response: *(a)* **without feedback** and *(b)* **with inverse feedback.**

The contrast is unmistakably in favor of the dual-gate MOSFET when GBW is of paramount importance. An additional benefit when using the dual-gate MOSFET is the simplicity of applying AGC by varying the bias on the second gate. Where this would severely modify the parasitic capacitances of a simple JFET, the dual-gate MOSFET, because of its greatly reduced feedback capacitance (and thus lower Miller effect), maintains a near-constant capacitance, thus reducing any detuning effects that would naturally occur with AGC control on the simple JFET.

Gain-bandwidth product can also be improved by the use of inverse feedback that is a maximum at band center and drops off toward the edges of the band. As shown in Fig. 4-11, a flattening of the response curve of the amplifier with a consequent increase in the bandwidth results. If the increase in bandwidth is greater than the decrease in gain, the GBW is increased.

4-7-5 The Broadband Amplifier

Undoubtedly the fundamental shortcoming of FETs as amplifiers has been their high output resistance (low output conductance), which has limited their application in broadband applications. The video source-follower amplifier is the notable exception since, being a *source* follower, its output impedance is effectively $1/g_{fs}$ in parallel with the source resistor R_s. However, the other amplifier configurations, common source, common gate, and Cascode, are generally plagued with a high drain impedance.

Yet there are solutions. Perhaps the simplest that offers increased high-frequency response, particularly in video amplifier applications, is the shunt peaking technique, where a small inductor is placed in series with the load resistor. The value of inductance for maximum flat-gain response may be calculated

$$L = \frac{R_1^2(C_{in} + C_{out})}{2} \tag{4-51}$$

Generally broadband techniques, other than the shunt-peaking method, involve Q manipulation of the output tuned circuit. This becomes especially critical when multistage amplifiers are built simply because of the bandwidth shrinking effect common to *synchronously tuned* stages:

$$BW_{overall} = BW_{one\,stage}[\sqrt{(2^{1/n} - 1)}] \qquad (4\text{-}52)$$

Perhaps the easiest and most popular method of improving bandwidth of the common-gate amplifier is to load the output to reduce Q. Once the overall gain and bandwidth have been established, the number of stages can be closely approximated and calculations made to determine the proper loading.

4-7-6 Amplifier Design Example

The following example should be of benefit to the reader in understanding the procedure.

Let an amplifier be designed for a 225-MHz center frequency with a 1-dB bandwidth of 50 MHz and 24 dB power gain using the Siliconix U310 JFET. Some familiarity with the U310 suggests that three stages are necessary. Solving for a single-stage bandwidth using the bandwidth shrinking formula, Eq. (4-52), the single-stage bandwidth for 50 MHz, 1 dB passband is 98 MHz. To achieve this bandwidth requires a loaded tuned circuit in the drain of each FET. The required Q is determined by

$$\frac{BW}{f_0} = \frac{1}{Q} \sqrt{\left(\frac{E_0}{E}\right)^2 - 1} \qquad (4\text{-}53)$$

Solving for Q gives

$$Q = 1.15$$

Referring to the admittance parameters of the U310 for a center frequency of 225 MHz, a close estimate of the drain load admittance equates to a drain output impedance of 3700 Ω shunted by approximately 2.5 pF (includes some stray capacitance). To achieve the required *single-stage* bandwidth,

$$Q = \omega C R_t \qquad (4\text{-}54)$$

where R_t is that value of load impedance required to achieve the single-stage bandwidth.

$$R_t = \frac{1.15}{1.415 \times 10^9 \times 2.5 \times 10^{-12}} = 330 \ \Omega \qquad (4\text{-}55)$$

Since the JFET itself exhibits a drain load resistance of 3700 Ω, the net resistance that must be placed across the tank circuit is about 365

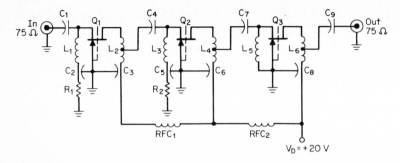

FIGURE 4-12 Schematic of a three-stage synchronously tuned bandpass amplifier ($C_1,C_4,C_7,C_9 = 68$ pF; $C_2,C_5 = 500$ pF; $C_3,C_6,C_8 = 1000$ pF; $Q_1,Q_2,Q_3 =$ Siliconix U310; $L_1,L_3,L_5 = 120$ nH; $L_2,L_4,L_6 = 222$ nH; $RFC_1,RFC_2 = 2.2$ μH; $R_1,R_2 = 51$ Ω).

Ω. This, of course, can be made a part of the tank circuit design. The circuit shown in Fig. 4-12 exhibits the response shown in Fig. 4-13.

Although this procedure (and those to follow) suggests that bandwidths can be achieved with FETs, one must not forget the limiting gain-bandwidth product (GBW), which remains inviolable. Consequently, there is a maximum bandwidth possible at a given overall gain. For synchronously tuned stages, shown in the example, maximum bandwidth occurs when the stage gain reduces to 4.34 dB (10 log e).

Other broadbanding schemes are available, and most are affected in some manner by bandwidth shrinking when multiple stages are used. It is, however, possible to reduce or even to eliminate bandwidth shrinkage by using over- and undercoupled multituned circuits, which have been well covered in the literature.

Another less popular method of improving bandwidth is the distributed amplifier. Early in the CATV era this had some popularity, but now it appears to be relegated to a few wideband oscilloscopes. In the distributed amplifier, rather than terminating the drain in a tank circuit, the parasitic capacitances of the FET contribute to the necessary shunting

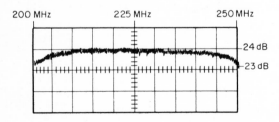

FIGURE 4-13 Passband response of a syncronously tuned bandpass amplifier.

transmission-line capacitance of the common-drain output network. Wide bandwidths are possible; however, the complexity of the design negates its popularity. One of the most debilitating effects for GBW is the gate conductance at high frequencies [or quality factor—see Eq. (4-5)].

4-8 MIXERS

There has been a great deal of interest in JFET mixers because they offer the opportunity of both high dynamic range and conversion gain. At one time it was thought that lower noise figures could also be attained, but experience has shown that with commercially available JFETs a noise figure comparable to that of the Schottky-barrier diode is, at best, difficult.

Perhaps the most favorable aspect of mixer design using JFETs is the square-law transfer characteristic. The JFET mixer has lower intermodulation and harmonic products than does a comparable bipolar or diode mixer. Active JFET mixers operating at high levels outperform passive mixers in terms of dynamic range and large signal-handling capabilities.*

Additionally, the active mixers offer improved conversion efficiencies over their passive counterparts. This permits relaxation of the IF amplifier gain requirements and in some cases the elimination of the RF preamplifier. The basic large-signal model, shown in Fig. 4-14, is considered to have an ideal square-law transfer characteristic, with a parasitic resistor R_s, in series with the source. The shunting capacitances are shown to be equal, as most FETs are closely symmetrical with respect to their gate.

To incorporate the effect of noise into the FET model, it is necessary to identify the sources of noise. Van der Ziel has shown that the primary sources of noise in a FET are bulk resistance in series with the drain and source, and channel-width modulation by thermal noise in the conducting channel; thus

$$\overline{i_n^2} = 4KTB\gamma g_{fs} \tag{4-56}$$

where $\gamma = 0.6$

A mixer circuit accepts a high-frequency modulated input signal, mixes it with a high-frequency unmodulated signal from a local oscillator, and produces a series of frequencies which are functions of the original

---

* It is also true that Cascoding diode sets has produced outstanding dynamic range and signal handling, but at the expense of vastly increased local oscillator power.

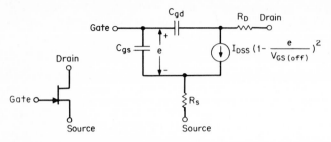

FIGURE 4-14 FET large-signal model ($C_{gs} = C_{gd} = C_o/[1 + (d/V)]^\alpha$; $C_{gs} = C_o/(d + V_{gs})^\alpha$; $C_{gd} = C_o/(d + V_{gd})^\alpha$; $\alpha = $ **law of capacitance variation for junctions**; $d = $ **work voltage for junction, usually 0.75 V**; $C_o = $ **junction capacitance at 0 V**; $I_{DDS} = $ **drain current for $e = 0$**; $V_R = $ **gate-to-source voltage for drain-$_{GS(off)}$ current equal to 1 μA). (Reprinted with permission from D. M. Hodson, "UHF Fet Mixer of High Dynamic Range," U.S. Army ECOM Research and Development Technical Report #ECOM-0503-P005-G821, December 1969.)**

two. One of these, usually the difference frequency, is selected by tuned circuits and is referred to as the "intermediate frequency" (IF). The IF, which is modulated with the same waveform as the incoming signal, is then amplified and the modulation extracted for whatever use the system requires. Mixers are generally used for frequency conversion, generally to a lower frequency for more effective handling, but quite often to higher frequencies where filtering can be more effective in removing unwanted sidebands (or images).

Initial evaluation of the active JFET mixer compared to bipolar mixers will imply a disadvantage because of local oscillator drive requirements; bipolar devices in low-level mixers require very little drive power. However, in high-level mixing this disadvantage is overcome because drive requirements are generally about equal.

Since JFETs have transfer characteristics approximating a square-law response, their third-order intermodulation distortion products are generally much smaller than those of bipolar transistors. Harmonic distortion and cross-modulation effects are third-order-dependent, and thus are greatly reduced when FETs are used in active balanced mixers.

A secondary advantage derives from available conversion gain, so that the FET mixer becomes simultaneously equivalent to both a converter and a preamplifier.

Design criteria, in order of priority, include the following:

1. Intermodulation and cross modulation

2. Conversion gain

3. Noise figure

4. Selecting the proper FET

5. Local oscillator injection

6. Designing the input transformer

7. Designing the IF network

4-8-1 Intermodulation and Cross Modulation

A basic aim in mixer design is to avoid the effects of intermodulation product distortion and cross modulation. Part of the problem may be resolved by using a balanced mixer circuit.

The active transfer function of the FET is represented by a voltage-controlled current source. For both cross modulation and intermodulation, the amount of distortion is proportional to the amplitude of the gate-source voltage. Since input power is proportional to input voltage and inversely proportional to input impedance, the best FET IM and cross-modulation performance is obtained in the common-gate configuration, where the impedance is lowest.

When JFETs are used as active mixer elements, it is important that the devices be operated in their square-law region. Operation in the FET square-law region will occur with the device in the depletion mode. Considerable distortion will result if the JFET is operated in the enhancement mode (positive V_{GS} for an n-channel JFET). The problems encountered are similar to those which arise when positive drive is placed on the grid of a vacuum tube.

Square-law-region operation emphasizes the importance of establishing proper drive levels for both quiescent bias and the local oscillator. The maximum conversion transconductance g_c is achieved at about 80 percent of the FET gate cutoff voltage $V_{GS(off)}$, and amounts to about 25 percent of the forward transconductance g_{fs} of the FET when used as an amplifier. Since conversion gain (or loss) must be considered, it is common to equate voltage gain A_V as

$$A_V = g_c R_L \tag{4-57}$$

where g_c is the conversion transconductance and R_L is the FET drain load.

An attempt to achieve maximum conversion gain by indiscriminately increasing the drain-load resistance will adversely affect any design priority concerning distortion—particularly intermodulation-product distortion.

Distortion takes different forms in mixers. Most obvious is that distortion which will occur if the FET is driven into the enhancement mode, as noted earlier. Finally, there is the so-called varactor effect.

The most frequent cause of poor mixer performance stems from signal overloading in the drain circuit. Excessive drain-load impedance degrades the intermodulation characteristics and produces unwanted cross-modulation signals.

A characteristic of the FET balanced mixer is that the correct drain-load impedance is inversely proportional to the value of the conversion transconductance. Figure 4-15 shows the improvement in IM characteristics obtained in the prototype mixer with the drain-load impedance reduced to 1700 Ω from 5000 Ω. Specifically, the dynamic load line must be plotted so that the signal peaks of the instantaneous peak-to-peak output voltage are not permitted to enter into the nonsaturated region of the FET. Suitable and unsuitable drain-load lines are shown in Fig. 4-16. Load impedance selection is quantified in Eqs. (4-72) through (4-74).

Distortion from the varactor effect is of secondary importance; it arises from an excessive peak voltage signal swing, where the changing drain-to-source voltage can cause a change in parasitic capacitance C_{rss} and give rise to harmonics.

A FET tends to be voltage-dependent when the drain voltage falls appreciably below 6 V. If the source voltage (from the power supply) is also low and the drain-load impedance is high, then distortion will develop. However, if proper steps are taken to prevent drain-load distortion, the varactor effect will also be inhibited.

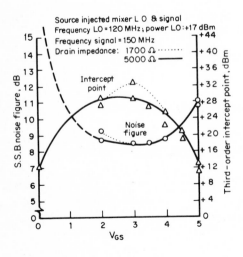

FIGURE 4-15 Comparison of mixer IM characteristics.

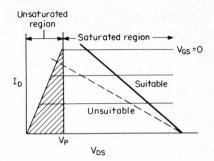

FIGURE 4-16 Plotting drain load lines.

4-8-2 Conversion Gain

In a FET, forward transconductance is defined as

$$g_{fs} = \frac{dI_D}{dV_{GS}} \tag{4-58}$$

and conversion transconductance is defined as

$$g_c = \frac{dI_{D(\omega i)}}{dV_{GS(\omega r)}} \tag{4-59}$$

where ωi = intermediate frequency
ωr = signal frequency

The effects of time-varying local oscillator voltage v_2 and the much smaller signal voltage v_1 must be considered:

$$v_{GS} = v_1 \cos \omega_1 t + v_2 \cos \omega_2 t \tag{4-60}$$

For square-law operation,

$$V_2 + V_{GS} \leqslant V_{GS(off)} \tag{4-61}$$

Drain current is approximately defined by

$$I_D = I_{DSS} \left(1 - \frac{V_{GS}}{V_{GS(off)}}\right)^2 \tag{4-62}$$

or

$$I_D \approx \frac{g_{fso} V_{GS(off)}}{2} \left(1 - \frac{v_{gs}}{V_{GS(off)}}\right)^2 \tag{4-63}$$

or

$$I_D \approx \frac{g_{fso}}{2 V_{GS(off)}} (V_{GS(off)} - v_{gs})^2 \tag{4-64}$$

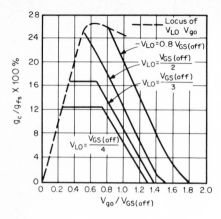

FIGURE 4-17 Normalized g_c/g_{fs} versus $V_{GS}/V_{GS(off)}$. (Reprinted with permission from S.-P. Kwok, "Field Effect Transistor RF Mixer Design Techniques," WESCON/67 *Convention Record*, Session 8.)

then

$$I_D \approx \frac{g_{fso}}{2V_{GS(off)}} \quad \text{(complex Taylor expansion)} \qquad (4\text{-}65)$$

which can be reduced to

$$I_{D(IF)} \approx \frac{g_{fso}}{2V_{GS(off)}} V_1 V_2 \cos(\omega_1 - \omega_2)t \qquad (4\text{-}66)$$

and the conversion transconductance is

$$g_c = \frac{g_{fso}}{2V_{GS(off)}} |V_2| \qquad (4\text{-}67)$$

Equation (4-67) suggests that g_c increases without limit as V_2 increases without limit. However, to avoid operation of the FET in the unsaturated region, the peak-to-peak swing of V_2 should not exceed $V_{GS(off)}$. Thus

$$2V_2 \text{ peak} \leqslant V_{GS(off)} \qquad (4\text{-}68)$$

or

$$V_2 \text{ peak} \leqslant \frac{V_{GS(off)}}{2} \qquad (4\text{-}69)$$

Figure 4-17 shows plots of normalized conversion transconductance, g_c/g_{fs}, versus normalized quiescent bias, $V_{GS}/V_{GS(off)}$, for different oscillator injections.

4-8-3 Noise Figure

Like the common-gate FET amplifier, the common-gate FET balanced mixer is sensitive to generator resistance R_g. A change of a decade in R_g can produce a noise-figure variation of as much as 3 dB.

4-8-4 How to Select the Proper FET

Conversion efficiency is determined by conversion transconductance g_c, which in turn is directly related to zero-bias saturation current I_{DSS} and the gate cutoff voltage $V_{GS(off)}$:

$$g_c = \frac{I_{DSS}}{V_{GS(off)}^2} |V_2| \tag{4-70}$$

$$\approx \frac{g_{fso}}{2 V_{GS(off)}} |V_2| \tag{4-71}$$

Equation (4-71) appears to indicate that FETs with high I_{DSS} are to be preferred. However, I_{DSS} and $V_{GS(off)}$ are related, and Figs. 4-18a and b show that devices from a family selected for high I_{DSS} do *not* provide high conversion transconductance, but actually produce a lower value of g_c.

Basic considerations in selecting FETs for this application are gate cutoff voltage $V_{GS(off)}$ for good conversion transconductance, and zero-bias saturation current I_{DSS} for dynamic range. Among currently available devices, the U310 offers excellent performance in both categories.

There is, of course, the possibility that FET cost is a major consideration in evaluating the active balanced mixer approach—the familiar price/performance tradeoff. If this is the case, there are a number of other FETs which will provide suitable alternatives to the U310. Remember, however, that conversion transconductance g_c can never be more than 25 percent of forward transconductance. Thus, as tradeoff considerations begin, the first sacrifice to be made will be the degree of achievable conversion gain. Intermodulation performance will follow, with the third tradeoff being available noise figure. Table 4-1 lists a number of possible alternatives to the U310.

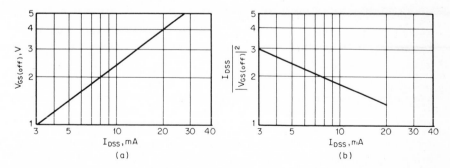

FIGURE 4-18 Relationship of I_{DSS} and $V_{GS(off)}$.

TABLE 4-1 COMPARISON OF JFET PARAMETERS CRITICAL TO MIXER PERFORMANCE

Typical characteristic	Device type		
	U310	2N4416	2N3823
g_m	14 kΩ	5 kΩ	3.5 kΩ
I_{DSS}	40 mA	10 mA	10 mA

4-8-5 Designing the IF Network

The IF network performs two important functions in the FET mixer circuit. It provides for optimum match between the FET and the IF amplifier at the IF frequency, and it effectively bypasses signal and local oscillator frequencies.

In network design, it is essential that the RF and local oscillator signals be sufficiently isolated from the intermediate-frequency signal to maintain rejection levels of at least 20 dB. If this isolation is not maintained, conversion gain and noise figure are degraded.

A first-order approximation to establish the proper load impedance may be obtained when

$$R_L = \frac{V_{DD} - 2V_{GS(off)}}{i_d} \qquad (4\text{-}72)$$

where

$$i_d = I_{DSS}\left(1 - \frac{v_{gs}}{V_{GS(off)}}\right)^2 \qquad (4\text{-}73)$$

and

$$v_{gs} = V_{GS} + V_1 \sin \omega_1 t \qquad (4\text{-}74)$$

For the U310 FET, the optimum drain-load impedance is established at slightly less than 2000 Ω, with sufficient local oscillator drive and gate bias determined from the conversion transconductance curve in Fig. 4-17.

4-9 BALANCED MIXERS

When high-performance, high-frequency junction FETs are used in active balanced mixers, the performance is superior to that obtained with hot-carrier diodes. Several types of mixers are compared in Table 4-2. The advantages and disadvantages of semiconductor devices currently used in various mixer circuits are shown in Table 4-3.

TABLE 4-2   COMPARISON OF MIXER TYPES

Characteristic	Mixer type		
	Single-ended	Single-balanced	Double-balanced
Bandwidth	Several decades possible	Decade	Decade
Relative IM density	1.0	0.5	0.25
Interport isolation	Little	10–20 dB	>30 dB
Relative LO power	0 dB	+3 dB	+6 dB

TABLE 4-3   COMPARISON OF SEMICONDUCTOR DEVICES FOR MIXERS

Device	Advantages	Disadvantages
Bipolar transistor	Low noise figure High gain Low dc power	High IM Easy overload Subject to burnout
Diode	Low noise figure High power handling High burnout level	High LO drive Interface to IF Conversion loss
JFET	Low noise figure Conversion gain Excellent IM products Square-law characteristic Excellent overload High burnout level	Optimum conversion gain not possible at optimum square-law response level High LO power
Dual-Gate MOSFET	Low IM distortion AGC Square-law characteristic	High noise figure Poor burnout level Unstable

4-9-1   Why FETs for Balanced Mixers?

Modern communication systems have the stringent requirements of wide dynamic range, suppressed of intermodulation products, and low cross modulation. These parameters must be considered before noise figure and gain are taken into account.

4-9-2   First-Order Single Balanced Mixer Theory

Essential details of balanced mixer operation, including signal conversion and local oscillator noise rejection, are best illustrated by signal-flow vector diagrams (Fig. 4-19).

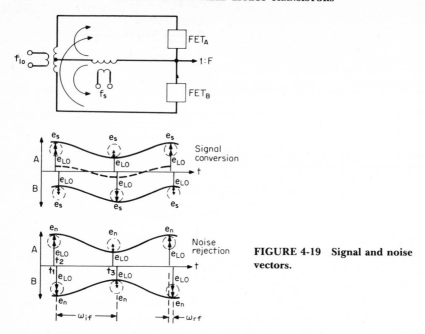

FIGURE 4-19   Signal and noise vectors.

Energy conversion into the intermediate-frequency (IF) passband is the major concern in mixer operation. In the following analysis, both the signal and noise vectors are shown progressing (rotating) at the IF rate (ω_{ift}); the resulting wave occurs through vector addition.

The analysis of local oscillator noise rejection (Fig. 4-19) assumes, for simplicity of explanation, that noise is coherent. Thus at some point in time t_1 the noise component e_n is "in phase" with the local oscillator vector e_{lo}, and FET A (the rectifying element) is ON; the JFET mixer acts as a switch, with the local oscillator acting as the switch drive signal. One-half cycle later, at time t_2, the signal flow is reversed for both the local oscillator vector and the noise component; FET A is OFF and FET B is ON. Moving ahead an additional one-half of the IF cycle, FET A is again ON, but the noise component has advanced 180° (ω_{ift}) through the coupling structure, and is now "out of phase." The process continually repeats itself.

The end result of this averaging (detection) is the cancellation of the noise which originated in the local oscillator, provided that the mixer balance is precise.

The analysis of the conversion of the signal to the IF passband is similar, but the signal is injected into the coupling structure at the equipotential tap. Thus at time t_2, the signal vector e_s is "out of phase"

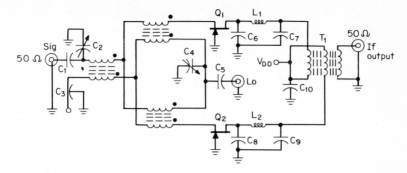

FIGURE 4-20 **Prototype active balanced mixer [C_1, C_5—0.01 μF; C_2, C_4—1 to 10 pF; C_3—1000 pF; C_6, C_8—30 pF; C_7, C_9—68 pF; C_{10}—0.1 μF; L_1, L_2—1.3 μH; Q_1, Q_2—U310 (2) or U430; T_1—RELCOM BT-9].**

with the local oscillator vector e_{lo}. The resulting envelope develops a cyclic progression at the IF rate, since the signal is "demodulated" by the mixing action of the FETs.

A schematic of a *prototype* balanced mixer is shown in Fig. 4-20. In the design of the prototype FET active balanced mixer, the generator resistance of the FETs is established by the hybrid coupling transformer. Two important criteria for the FETs in the circuit are high forward transconductance and a value of power-match source admittance g_{igs} which closely matches the output admittance of the coupling transformer. In the common-gate configuration, match points for optimum power gain and noise do not occur at the same value of generator resistance, as shown in Fig. 4-21. Optimum noise match can only be achieved at the sacrifice of bandwidth.

Best mixer performance is achieved with "matched pairs" of JFETs. Basic considerations in selecting FETs for this application are gate cutoff voltage $V_{GS(off)}$ for good conversion transconductance, and zero-bias saturation current I_{DSS} for dynamic range. A match to 10 percent is generally adequate. Among currently available devices, the Siliconix U310 and the dual U431 offer excellent performance in both categories; common-

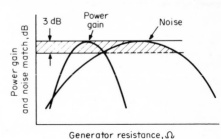

FIGURE 4-21 Power gain and noise matching.

gate forward transconductance is 20,000 μmhos max at $V_{DS} = 10$ V, $I_D = 10$ mA, and $f = 1$ kHz.

4-9-3 Local Oscillator Injection

Low IM distortion products and noise figure, plus best conversion gain, will be achieved if the voltage swing of the local oscillator across the gate-to-source junction is held to the values presented in Fig. 4-17. V_{LO} is expressed in terms of peak-to-peak voltage, while $V_{GS(off)}$ is a dc voltage.

Local oscillator injection can be made either through a brute-force drive into the JFET source through the hybrid input transformer, or through a direct-coupled circuit to the JFET gates, where less drive will be required for the desired voltage swing. Two circuits to obtain direct gate coupling are shown in Fig. 4-22.

The source-injection method is used in the design of the present mixer to maintain the inherent stability of a common-gate circuit. A minor disadvantage with the direct-drive method is that the required gate-to-source voltage swing requires considerable local-oscillator input power. For source injection through the transformer, best mixer performance is obtained with a local-oscillator drive level of +12 to +17 dBm across a 50-Ω load.

Direct coupling to the FET gates would occur at a higher impedance level, which would result in less local oscillator drive power. However, when the gates are tied together, shunt susceptance requires some form of conjugate matching. This brings about an undesirable reduction of instantaneous mixer bandwidth.

The performance of the active mixer is clearly superior to that of diode mixers, as shown in Fig. 4-23 and Table 4-4; it contributes to

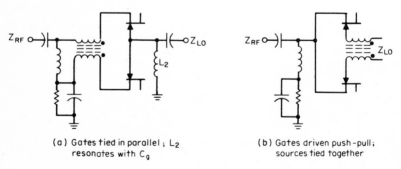

(a) Gates tied in parallel; L_2 resonates with C_g

(b) Gates driven push-pull; sources tied together

FIGURE 4-22 Alternate forms of LO injection.

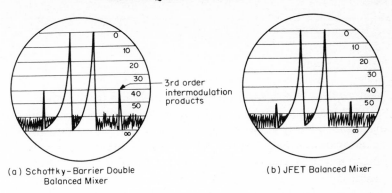

(a) Schottky-Barrier Double
Balanced Mixer

(b) JFET Balanced Mixer

FIGURE 4-23 Comparison of third-order IM products.

overall system gain in areas critical to telecommunications practice and reduces associated amplifier requirements.

4-10 DOUBLE-BALANCED MIXERS

The comparison in Table 4-2 clearly shows those performance characteristics of the double-balanced mixer which have made it the most popular of all mixer types. Among these attributes are greatly improved interport isolation and a significant degree of rejection of local oscillator carrier amplitude modulation.

Passive devices, however, such as Schottky-barrier (hot-carrier) diodes, have certain fundamental shortcomings, such as high conversion

TABLE 4-4   50 TO 250-MHz MIXER PERFORMANCE COMPARISON

Characteristic	JFET		Schottky		Bipolar	
Intermodulation intercept point	+32	dBm	+28	dBm	+12	dBm*
Dynamic range	100	dB	100	dB	80	dB*
Desensitization level (the level for an unwanted signal when the desired signal first experiences compression)	+8.5	dBm	+3	dBm	+1	dBm*
Conversion gain	+2.5	dB†	−6	dB	+18	dB
Single-sideband noise figure @ 50 MHz	7.2	dB	6.5	dB	6.0	dB

* Estimated.

† Conservative minimum.

loss and high local oscillator drive requirements. The active balanced mixer which employes FETs is a welcome innovation. Conversion gain and improved intermodulation distortion characteristics alone place the FET double-balanced mixer far ahead of its passive counterparts. The high saturation levels possible with modest local oscillator power make such a mixer useful for mixing both small and large signals.

Double-balanced mixers using MOSFETs have been considered; however, the MOSFETs were used solely as switching devices, requiring no external dc power. As a result, MOSFET mixers have exhibited high conversion loss and require considerable local oscillator drive power.

4-10-1   First-Order Double-Balanced Mixer Theory

In either single- or double-balanced mixer design, the prime requirement is that when the mixer is excited by the local oscillator carrier, the circuit must be capable of rejecting the amplitude-modulated wave which exists about the LO. Also, the mixer must reject any AM signal entering from the local oscillator port. (This signal rejection is usually known as AM local oscillator noise cancellation.)

A second requirement for balanced mixers is the establishment of interport isolation between the signal, local oscillator, and IF ports. A third desirable characteristic is the reduction of intermodulation distortion products. Careful attention to the design of double-balanced mixers will satisfy the foregoing criteria.

Figure 4-24 shows the schematic of a JFET double-balanced mixer. The four high-performance junction FETs are chosen for closely matched characteristics. The significance of the quad-FET configuration will be dealt with later.

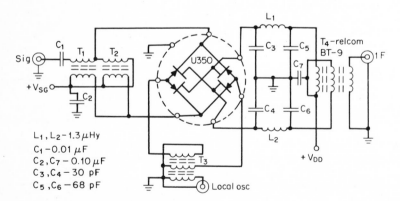

FIGURE 4-24   Double-balanced mixer.

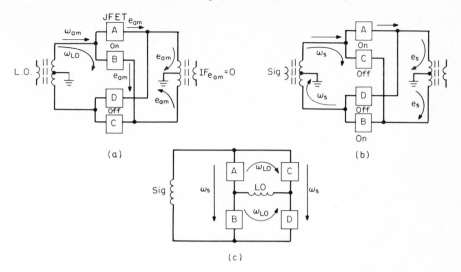

FIGURE 4-25 Double-balanced mixer analysis.

The schematic of Fig. 4-25*a* is simplified to show only the local oscillator circuit so that the rejection mechanism of AM signals, either on the LO carrier or entering through the local oscillator port, is easier to understand. Likewise, the equivalent circuit in Fig. 4-25*b* demonstrates how the signal is enhanced at the IF output. Local oscillator AM cancellation and signal enhancement are dependent upon the precise balance of the IF transformer, as well as on the match of the four FETs which make up the quad network. In Figure 4-25*c*, the schematic has been rearranged to show both the local oscillator and the signal input transformers; the mechanics of interport isolation may be easily visualized. Signal excitation provides an equipotential at the junctions of the local oscillator transformer and FET pairs *AB* and *CD;* in the same manner, excitation of the local oscillator produces an equipotential balance at the junctions of the signal transformer and FET pairs *AC* and *BD.*

Harmonic distortion products are reduced by the balance between the signal and local oscillator (inputs) and the IF (output), where even-integer harmonics of the signal and local oscillator frequencies are effectively cancelled. A sixth-order summary of such products in both single- and double-balanced mixers is shown in Table 4-5. Note how the relative densities agree with Table 4-2. The effects of harmonic distortion can be reduced by a judicious selection of the IF passband response. Third-order IMD (intermodulation distortion) products are reduced by virtue of the characteristics of junction FETs, which approximate a square-law response. Care must be taken in FET operation, however, to avoid

TABLE 4-5 COMPARISON OF MODULATION PRODUCTS IN SINGLE- AND DOUBLE- BALANCED MIXERS TO SIXTH ORDER

Single-balanced	Double-balanced
f_s	
$3f_s$	
$5f_s$	
$f_0 \pm f_s$	$f_0 \pm f_s$
$f_0 \pm 3f_s$	$f_0 \pm 3f_s$
$f_0 \pm 5f_s$	$f_0 \pm 5f_s$
$2f_0 \pm f_s$	
$2f_0 \pm 3f_s$	
$3f_0 \pm f_s$	$3f_0 \pm f_s$
$3f_0 \pm 3f_s$	$3f_0 \pm 3f_s$
$4f_0 \pm f_s$	
$5f_0 \pm f_s$	$5f_0 \pm f_s$

driving the device into forward conductance by the application of too much local oscillator power.

4-10-2   Harmonic Distortion, Intermodulation Products, and Cross Modulation

Spurious output signals in mixers fall into three categories:

1. Spurious mixer products derived from harmonic mixing of the signal and local oscillator frequencies

2. m-Tone, n-order intermodulation products

3. "Chirping," which arises from undesired mixing frequencies falling in the IF passband

The harmonics of a single-signal frequency, when mixed with the harmonics of the local oscillator, produce spurious outputs which are level-dependent on the signal amplitude. These products are greatly reduced by the double-balanced mixer, where the even harmonics are effectively cancelled; when FETs are used, the Taylor-series power expansion falls quickly to zero above the second order.

However, modulation products of a similar nature will arise if the broadband down-converting mixer is not preceded by signal preselection, because of the mixer's equal response to the "image" frequency. Here, perfectly valid signals will mix with the local oscillator, producing interfering IF signals whose only difference, when compared to the desired IF signal, is that they move counter to the desired IF signal when the local oscillator is shifted.

Two-tone, odd-order IM products differ markedly from other spurious signals. This form of harmonic distortion consists of interactions between two or more input signals and their respective harmonics. In turn, these products are mixed with the fundamental and harmonics of the local oscillator, generating spurious products which may fall within the IF passband, on or very near to the desired signal.

Cross modulation in the active JFET balanced mixer does not pose a serious problem provided the signal input is maintained at a high conductance, which will occur with source injection. Cross modulation is very dependent on and directly related to the impedance across which the signal is impressed. In the active JFET double-balanced mixer this impedance is very low, typically 35 Ω. Consequently, the effects of cross modulation may be disregarded.

In the mixing process of any active device, the value of the FET drain current may be derived from a knowledge of the transconductance of the device and the impressed signal voltage e_g. This is obtained from the Taylor-series power expansion:

$$i_d = g_m e_g + \frac{1}{2!}\frac{\partial g_m}{\partial V_G}e_g^2 + \frac{1}{3!}\frac{\partial^2 g_m}{\partial V_G^2}e_g^3 + \cdots + \frac{1}{n!}\frac{\partial^{n-1}g_m}{\partial V_G^{n-1}}e_g^n \qquad (4\text{-}75)$$

which can be broken down into the components shown in Table 4-6.

In FET theory, the second- and higher-order derivatives of g_m are absent, and the device thus offers a considerable reduction of both intermodulation products and higher-order harmonics. In the double-balanced mixer, where $F1 \pm F2$ is the desired result, it is well to manipulate mixer design and bias conditions to render $\partial g_m/\partial V_G$ as large as possible, simultaneously reducing all other terms.

TABLE 4-6   ELEMENTS OF THE TAYLOR POWER SERIES

Term	Output	Transfer characteristic
$g_m e_g$	F1, F2	Linear
$\dfrac{1}{2!}\dfrac{\partial g_m}{\partial V_G}e_g^2$	2F1, 2F2 F1 $\pm$ F2	Second-order square-law
$\dfrac{1}{3!}\dfrac{\partial^2 g_m}{\partial V_G^2}e_g^3$	3F1, 3F2 2F1 $\pm$ F2 2F2 $\pm$ F1	Third-order

4-10-3 Local Oscillator Injection

Local oscillator drive for active FET mixers, either balanced or unbalanced, differs from the drive characteristics of passive diode mixers. In the switching mode, the diode mixer requires sufficient local oscillator drive to swing the diodes from a hard ON state to a hard OFF state. For best IMD performance, the gate of the FET must never be driven positive with respect to the source—a case equivalent to the hard ON condition of the diode. Consequently, local oscillator drive for the balanced mixer is less than that required for a passive balanced mixer *with comparable performance characteristics.*

The double-balanced mixer relies on balanced drive from both the local oscillator and the signal source. Since conversion efficiency, optimum noise figure, and good cross-modulation effects can best be served with the signal entering through the common quad JFET source, the local oscillator excitation may be applied directly at the gates of the FET array.

4-10-4 AM Local Oscillator Noise Rejection

Originally, balanced mixers were used for the specific purpose of cancelling spurious AM signals existing on or about the local oscillator carrier (the function of the mixer in establishing good interport isolation was a side effect). These signals could be either spurious AM signals generated on or about the carrier (Fig. 4-26) or actual signals existing at the signal frequency. In the latter case, the signals enter the mixer through the local oscillator, having found their way in through some leakage coupling phenomenon.

Regardless of the type or source of AM signals entering through the local oscillator port, the balanced mixer should effectively reject these signals so that their products do not occur at the intermediate

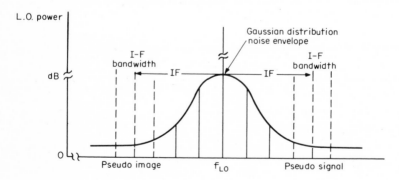

FIGURE 4-26 Generation of spurious AM signals.

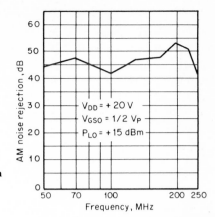

FIGURE 4-27 AM noise rejection in double-balanced mixer.

frequency. In the early days of balanced mixers, a 20-dB rejection of AM noise was considered good; today's sophisticated techniques for selection of dynamically matched semiconductors can provide AM rejection in excess of 30 dB. Figure 4-27 provides an insight into the degree of AM noise rejection available in the double-balanced mixer. (Insofar as FM noise is concerned, it should be noted that no mixer is capable of rejecting frequency-modulated signals entering through the local oscillator.)

An interesting point not generally considered in discussions of balanced mixers is that the dynamic range of the mixer is limited by the conversion of local oscillator noise into the intermediate frequency. This blanks out a weak signal and places a bottom on sensitivity.

4-10-5 Interport Isolation

Like AM noise rejection and dynamic unbalance, interport isolation is very dependent on mixer balance (symmetry). Matching aspects of the JFET quad array and the phase/amplitude balance of the signal input and local oscillator input transformers play important roles in achieving interport isolation. Capacitive and magnetic coupling between the transformers add to problems of interport isolation in balanced mixers.

Interport isolation was enhanced in the prototype mixer through careful parts layout. As a measure of the overall effects of unbalance, a quantitative measurement of interport isolation vs. dynamic unbalance is made in Fig. 4-28. Dynamic unbalance may be regarded as another expression for AM noise rejection, except that the latter does not provide a ready insight into the effects of symmetry, balance, and quad match.

In Fig. 4-29, the interport isolation between the local oscillator and signal input ports is shown to be 35 dB typically.

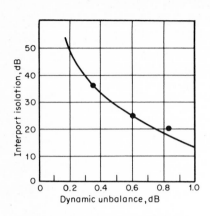

FIGURE 4-28 Interport isolation vs. dynamic unbalance.

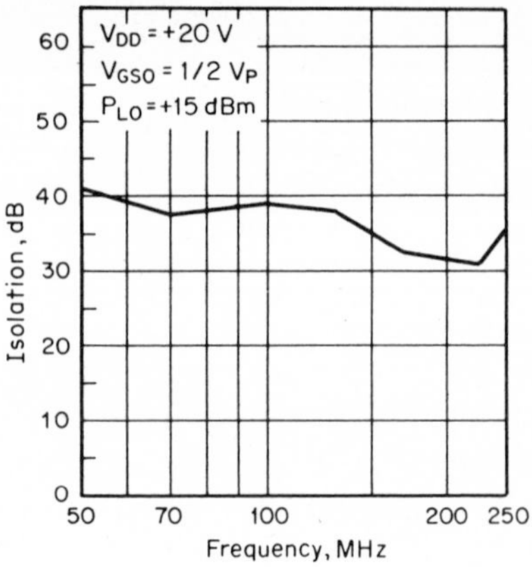

FIGURE 4-29 Interport isolation.

Selection of the dynamic drain impedance in the IF network is a critical point in the design. Both IM product distortion and cross modulation will be affected by the instantaneous peak-to-peak voltage of the FETs if the dynamic drain impedance allows the signal peaks to enter either the pinchoff or breakdown-voltage regions of the transistors. Here another design tradeoff must be considered. If the impedance is too high, the dynamic range of the mixer will be limited; if the impedance is too low, useful conversion gain will be sacrificed, as shown in Fig. 4-30.

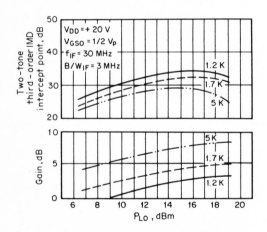

FIGURE 4-30 Gain and IMD vs. local oscillator drive.

4-10-6 Mixer Performance

Quad FET arrays with both high and low pinchoff-voltage levels were used in evaluation of the active double-balanced mixer; the mixer exhibited clearly superior characteristics, compared to equivalent small-signal passive double-balanced mixers. The low- to medium-pinchoff-voltage FET array performed slightly better than the high-pinchoff devices solely because of a limitation in available local oscillator power.

4-11 OSCILLATORS

FETs have some fundamental advantages over bipolar transistors in oscillators. An outstanding advantage is their very low sideband noise. In addition, they may be designed so that the frequency of oscillation is relatively independent of bias current. This reduces the effects of drift during temperature fluctuations. The parasitic capacitances of FETs are known to be voltage-dependent, and this can be turned to great advantage since it makes possible easily designed VTOs (voltage-tuned oscillators).

Although noise is often difficult to characterize because of its random or nondeterministic nature, it is possible to differentiate various forms of noise through an understanding of the gaussian distribution of noise about an RF carrier. The three major forms of noise are (1) low-frequency noise ($1/f$), (2) thermal noise ($4kTBR$), and (3) shot noise ($\bar{i}_n$). These types of noise can be identified from their relationship to the main RF carrier. For example, low-frequency noise predominates very close to the carrier and falls to insignificant levels when it is displaced more than 250 Hz from the carrier. Thermal noise plays the dominant role in the midfrequency region from the $1/f$ noise decay point to approximately 20 kHz from the carrier. It is commonly associated with equivalent resistance where the rms value of noise voltage of the Thevenin generator becomes the classic $\sqrt{4kTBR}$. Noise appearing beyond 20 kHz is known as shot noise, and is directly attributed to noise current. Because of the typically uniform distribution of shot noise, it is often referred to as "white noise."

Generally oscillators are represented as narrowband circuits; consequently most interest in noise centers about a reasonably narrow band. The voltage that results when white noise is passed through any relatively narrow bandpass may be described by

$$v(t) = x_1(t) \cos \omega_{ot} + x_2(t) \sin \omega_{ot} \qquad (4\text{-}76)$$

where ω_o is the midband angular frequency and $x_1(t)$ and $x_2(t)$ are uncorrelated functions of time which vary slowly and randomly about zero in a manner described by the gaussian probability density.

4-11-1 Origins of Oscillator AM Noise

Although an oscillator tends to produce a wave that is nearly sinusoidal, there are other fluctuations present. When the energy in the frequency domain close to the carrier is observed on a spectrum analyzer, noise appears as a modulation phenomenon. This observation would be greatly enhanced if the noise contribution were coherent and consisted of discrete sideband frequencies. The major component of AM noise is low-frequency noise $(1/f)$. Both thermal and shot noise are relatively insignificant segments of AM noise compared with $1/f$. A graph of AM noise vs. frequency removed from the carrier is shown in Fig. 4-31.

4-11-2 Design of a VHF Oscillator

Important design considerations for best oscillator performance include using a FET with high forward transconductance (g_{fs}), maintaining the gate at ground potential, and keeping a high unloaded tank Q. The g_{fs} reduces the effective noise resistance. The grounded gate reduces the noise voltage resulting from the product of the gate leakage current and the series gate resistance. The high tank-circuit Q serves as an effective filter for the sideband noise energy.

The example used to illustrate an ultra-low-noise design is somewhat extraordinary for a circuit employing a FET. The FET chosen was the Siliconix U310, which has a forward transconductance value higher than 18 mmhos at zero bias $(V_{GS} = 0)$. The oscillator basically consists of two coaxial resonators, one for the FET source and the other for the drain. Oscillation is established by capacity coupling between the two resonators; output coupling is derived from the magnetic coupling which

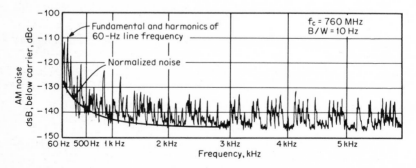

FIGURE 4-31 AM noise vs. frequency removed from the carrier.

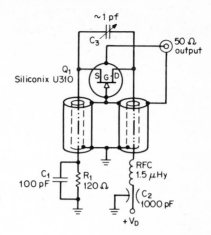

FIGURE 4-32   Oscillator circuit.

exists at the open ends of the resonators. Optimum resonator Q is achieved by designing the coaxial resonators for a characteristic impedance of 75 Ω. The oscillator circuit is shown in Fig. 4-32.

The technique to establish the proper resonator length for the desired frequency requires a first-order approximation of the anticipated capacitive fringing which derives from both the FET and the feedback network. A short-circuited coaxial transmission line is theoretically resonant at a quarter-wavelength of the resonating frequency, except for the effects of fringe field capacitance. At resonance,

$$X_L = X_C \tag{4-77}$$

If the fringe capacitance is known, X_C can be calculated as

$$X_C = \frac{1}{\omega C} \tag{4-78}$$

From this, the resonator length can be determined as

$$X_C = \tan \beta l \tag{4-79}$$

In making these calculations, a Smith chart is invaluable, as is shown in the following illustration:

$$
\begin{array}{ll}
\text{Frequency of oscillation} & = 760 \text{ MHz} \\
\text{FET } b_{igs} \text{ (from data sheet)} & = 16 \text{ mmhos} \\
\text{Capacitance from } b_{igs}\text{:} & C_{gs} = 3.4 \text{ pF} \\
\text{Allow for stray capacitance} & \\
\quad \text{and the feedback network:} & C_s = \underline{1.5 \text{ pF}} \\
& \quad\quad 4.9 \text{ pF}
\end{array}
$$

Thus $X_C = j0.57$ (normalized to 75 Ω)

Locate 0.57 on the Smith chart. The wavelength toward the load = 0.081λ. Since a wavelength at 760 MHz is 39.5 cm, the resonator cavity length is simply

$$39.5 \times 0.081 = 3.20 \text{ cm } (1.26 \text{ in}) \tag{4-80}$$

In the completed FET coaxial oscillator circuit, the output-coupling loop consists of a single turn made fast to the cavity by the BNC flange and the FET itself.

4-11-3 Conclusions

Measured performance of the oscillator is shown in Table 4-7A; AM noise measurements in a 10-Hz bandwidth are shown in Table 4-7B.

TABLE 4-7A   OSCILLATOR MEASURED PERFORMANCE AT 25°C

V_{DD} (V)	+10	+15	+20	+25
I_D (mA)	15	16.2	18.2	21
P_{out} (dBm)	+6.6	+15.2	+18.3	+20
Frequency (MHz)	725	742.7	754.7	762.9

TABLE 4-7B   AM NOISE MEASURMENT

Frequency displaced from carrier	dBc
50 Hz	−130
500 Hz	−139
1 kHz	−143.5
5 kHz	−146

The Reike diagram shown in Figure 4-33 makes possible the accurate prediction of expected power output and operating frequency and the oscillator feeding directly into a mismatched load. Expansion of the Reike diagram to show frequency vs. transmission-line length (in degrees) will allow prediction of the long-line effect on oscillator stability.

4-12   THE HIGH-FREQUENCY POWER FET

In the early 1960s a fever spread among many laboratories throughout the world to develop a high-frequency power FET, but although many attempts were made, none really ever became successful to the point of offering a commercial product. The first attempt is believed to have

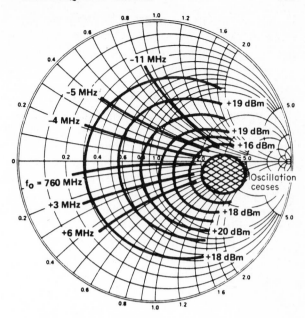

FIGURE 4-33   Reike diagram.

been in the U.K. Soon after, however, Teszner of France made some outstanding contributions in FET design and predicted high-frequency performance. In the mid-1960s, he coauthored a paper on the "Gridistor," which promised high-frequency performance. The U.S. Signal Corp financed R.C.A. in the development of the first successful RF MOSFET. An output of 14 W at 10 MHz was reported. About the same time (1968) Japanese researchers published a brief report describing a radically new structure called "VMOS." Although it was not specifically mentioned as having high-frequency capabilities, this VMOS structure has become one of the dominant technologies for high-frequency power. Soon after reporting on this VMOS, the Japanese followed with their "static induction transistor" using the modified JFET approach of Teszner, Zuleeg, and Nishizawa. They have made some very impressive gains in offering power through the UHF spectrum; however, this work has been limited to the laboratory at the time of this writing. In mid-1973 the Russians announced a high-frequency power MOSFET, the КП901А series, purported to be commercially available and offering upwards of 10 W through 100 MHz.

Since about 1970 DMOS (double-diffused MOS) has been the predominant technology in high-frequency MOS, and is slowly finding application in medium-power RF designs; VMOS has taken a very definite lead in high-frequency power. VMOS possesses all the attributes of any majority-carrier FET; when it is used as a power device at high frequency, these advantages manifest themselves by giving it the ability to withstand any mismatch without fear of burn-out.

An additional advantage of VMOS at high frequency is its higher input, or gate, impedance. Typically it is an order of magnitude greater than an equivalent-powered high-frequency bipolar. A higher input impedance allows easier broadband matching. The V-MOSFET also offers outstanding forward gain and high reverse isolation, almost to the point of offering truly unilateral amplifiers when properly neutralized.

Being a majority-carrier transistor, VMOS offers exceptionally high-speed switching characteristics and becomes especially suited for high-class amplifier service, such as class D, E, or F, where very high efficiencies are possible.

Unlike many high-frequency bipolar transistors, VMOS also offers outstandingly low noise both as a small-signal RF amplifier and, more importantly, as a power amplifier. A broadband VMOS amplifier contains far less AM sideband noise than does any power-comparable bipolar transistor.

Figure 4-34 illustrates the simplicity of designing a broadband amplifier. A Siliconix type VMP4 is the VMOS device. Although not well matched to the 50-Ω input (a VSWR running to 2.5:1), the balun transformer represents an easy solution for a near decade bandwidth amplifier.

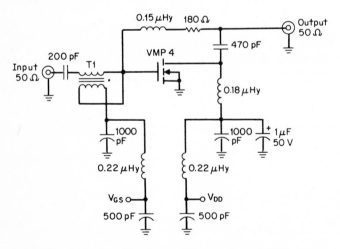

FIGURE 4-34 Broadband amplifier using VMOS power transistor.

The output circuitry is more conventional, based on the formula established for most power designs:

$$R_L = \frac{(V_{DD} - V_{SAT})^2}{2P} \tag{4-81}$$

4-12-1 Power FETs in Switch-Mode Amplifiers

VMOS has many advantages over its bipolar counterpart for both class B and clsss D RF power amplification. In particular,

1. The high input impedance and the negative temperature coefficient of the FET make complicated bias supply unnecessary.

2. Drive power is reduced by the high input impedance.

3. Input matching is simplified by the high input impedance of the gate.

4. Second breakdown is inhibited because of the negative temperature characteristic; hence the FETs can withstand reactive loads in class B operation.

5. The absence of storage time facilitates rapid switching in class D service.

6. The absence of storage time eliminates the possibility of subharmonic oscillation in either class D or saturated class B operation.

7. The ability of MOSFETs to pass current in both directions allows them to operate in class D with reactive loads without diode protection. This is especially significant since such diodes are not generally available.

8. The saturation resistance of a FET (in contrast to the saturation voltage of a bipolar transistor) allows the efficiency in class D operation to remain high even at low power levels. This is significant since SSB voice signals consist mainly of low amplitudes.

9. The low gate-drain capacitance reduces feedback, making the amplifier easy to stabilize.

10. The low gate-drain capacitance also reduces feedthrough. This improves linearity when supply-voltage modulation is used in class D service.

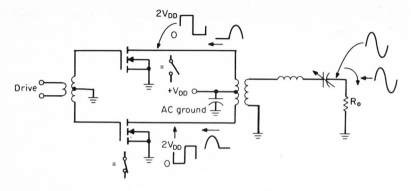

FIGURE 4-35 Class D amplification—FETs are switches.

11. The essentially constant gain of the VMOS over the entire HF band eliminates the need for gain-leveling circuitry.

Class D switching amplifiers are controlled by the driving signal to cause the FETs to act as switches. The voltage-switching configuration shown in Fig. 4-35 has a square-wave drain voltage produced by the alternate switching of the two FETs. The series-tuned output circuit passes current only at the fundamental (switching) frequency, and the sinusoidal output voltage is equal to the fundamental frequency component of the square wave on the transformer secondary winding. Alternate half-cycles of the transformed output current flow through each FET. The amplitude of the output is a function of the supply voltage and is not affected by the amplitude of the driving signal if it is sufficient to cause switching. Bias is useful simply to reduce the amplitude of the driving signal. A class D amplifier can ideally achieve 100 percent efficiency at all output levels; in practice, its efficiency is greater than that of a class B amplifier, especially at lower output levels.

Linear amplitude modulation is readily accomplished by variation of the drain-supply voltage. Single-sideband signals may be amplified through the envelope elimination and restoration technique, which uses amplitude modulation to restore the envelope of the SSB signal and a limiter to ensure adequate drive for the power amplifier.

4-13 NOISE TEMPERATURE AND NOISE FIGURE

Johnson noise, defined as a nonperiodic ac voltage fluctuation, places a limit on the sensitivity of amplifiers and mixers. All high-frequency transistors, from audio frequencies up, whether FETs or bipolars, as well as all passive elements, reach a well-defined limit of performance based on Johnson noise.

Historically, active transistors have been categorized by their *noise figure;* the lower the value, the better the device. Perhaps two decades ago, hidden within the confining technology of ultra-low-noise parametric amplifiers and hydrogen masers, was a unique term of performance called noise temperature. Now this "new" term noise temperature is emerging and finding increasingly wider acceptance in the trade.

Since Johnson noise is an effect of electron agitation, it is reasonable to assume that at absolute zero ($-273.18°C$) Brownian movement, and hence electron agitation, ceases. Consequently, since our standard operating temperature is generally regarded as $20°C$, the standard noise temperature T_0 of a passive device, such as a resistor, would be 293 K. By universal scientific consent, and for convenience in mathematical analysis, T_0 has been established at 290 K.

Again, since Johnson noise is electron agitation, it is temperature-dependent. However, it must be understood and appreciated that active components, such as transistors, can exhibit noise temperatures *differing* from their operating temperatures as well as their ambient temperatures! This exhibited noise temperature is the "effective input noise temperature" T_e of the active device.

Many readers may be aware of the term "input available noise power" as $P = kTB$, where k is Boltzman's constant, B is bandwidth in hertz, and T is temperature. The noise output of an active transistor is the sum of the input noise and the noise contributed by the device:

$$N_{\text{power}} = GkB(T_{\text{in}} + T_e) \tag{4-82}$$

where G is the gain and T_e is that term defined above as effective input noise temperature.

The definition of "noise figure" is the ratio of the total noise power delivered when the noise temperature at its input is 290 K to that of the input. In other words, by formula:

$$NF = \frac{N_{\text{power}}}{GkB(290)} \tag{4-83}$$

Combining Eqs. (4-82) and (4-83), we obtain

$$NF = 1 + \frac{T_e}{290} \tag{4-84}$$

Since NF is generally identified in decibels, the above formula (4-84) may be expanded to

$$NF_{\text{dB}} = 10 \log\left(1 + \frac{T_e}{290}\right) \tag{4-85}$$

or

$$T_e = 290 \left[\left(\text{antilog} \, \frac{\text{NF}}{10} \right) - 1 \right] \quad \text{K} \tag{4-86}$$

For example, the U310/NZA has a noise figure of 2.7 dB at 450 MHz. What is its noise temperature?

$$T_e = 250 \, \text{K}$$

4-13-1   Optimum Noise Matching of Amplifiers

In any amplifier or linear two-port, whether performing at audio or at some high frequency, any noise generated by the input stage tends to obscure or mask the amplified signal appearing at the output. This is so because noise is as readily amplified as signal. Consequently, it should be a designer's goal to minimize this input-contributed noise. To do so will greatly enhance the small-signal performance of the amplifier. Such an amplifier is said to be "noise matched."

Noise is generated by thermal agitation within resistive elements. Pure reactive elements (such as L and C) do not contribute noise. Consequently, when the total input admittance is a pure conductance, the noise figure of the amplifier is minimized—that is, when

$$j y_{\text{in}} = 0$$

From this, quite apart from bandwidth requirements, it is desirable to have the input circuit resonant at band center. Any losses within the input stage will increase the noise figure even if these losses have an effective noise temperature of zero!

It is known that the equivalent input noise resistance for JFETs is

$$R_{\text{eq}} = \frac{0.67}{g_{fs}} \tag{4-87}$$

It has been established that to obtain the lowest first-stage noise figure, the input admittance is a root function of the ratio of a complex transistor conductance G_b to its equivalent noise resistance R_{eq}; thus

$$G_{s(\text{opt})} = \sqrt{\frac{G_b}{R_{\text{eq}}}} \tag{4-88}$$

where G_b is the additive combination of both the shunt input conductance as a function of FET gate current I_g and the equivalent conductance of the gate-to-source as a function of frequency.

The best way to determine optimum noise source resistance for most amplifiers is to measure it! This is not always an easy task. Generally the optimum source resistance turns out not to be a resistance *per se*, but an optimum source admittance, that is, a complex quantity consisting of both conductance and susceptance, viz., $Y_{s(opt)} = G + jB$.

Optimum source admittance for FETs varies both with frequency and with configuration; common-source input stages generally have higher values of admittance than would the same FET in a common-gate configuration. Likewise, since gate current is critical to the optimum input noise admittance, the FET geometry plays an important role.

For the Siliconix U310 (NZA geometry), the optimum input source admittance for the *common-gate configuration* has been determined:

450 MHZ $\quad R_{eq} = 166 \; \Omega$ in parallel with $X_l = 120 \; \Omega$

4-13-2 Optimum Noise Bandwidth of Amplifiers

The bandwidth of an amplifier contributes to its faithfulness in reproduction as well as to its dynamic range. This does not mean that the wider the bandwidth, the better the fidelity, for such is not always the case. The bandwidth must be sufficiently broad to accept the total signal, yet not so wide that spurious and Johnson noise mask the weaker signals.

Dynamic range simply means that the amplifier not only detects the *weakest* possible signal but also will not readily overload under strong input signals. A steady-state amplifier has far different design requirements than a transient (fast-signal) amplifier. For the latter, dynamic range also suggests purity of response.

The detectability of weak signals involves many aspects; what will be addressed here is simply the noise bandwidth.

A narrowband amplifier will amplify only a narrow spectrum of signals. If the passband is less than the signal spectrum, the fidelity will be impaired because the output signal will reach only a fraction of its peak value. Since noise power is a linear function of bandwidth, a narrowband amplifier will not have the noise susceptibility of its wide-bandwidth counterpart. Even though the noise is restricted because of bandwidth, signal fidelity is also restricted *if* the signal spectrum exceeds the amplifier bandwidth.

As the passband is widened, the signals are reproduced more faithfully until the output reaches its peak value. Simultaneously, the noise power increases. It should be evident that eventually an optimum bandwidth occurs at which there is no further improvement in signal fidelity. Any additional increases in bandwidth only generate increasingly greater amounts of noise power that do nothing beneficial to the signal. In fact, on the contrary, small signals are masked, thus reducing the dynamic range of the amplifier.

An amplifier with too narrow a passband restricts the fidelity (as any audiophile is well aware), and too wide a passband contributes excessive noise, reducing the signal-to-noise ratio of the amplifier.

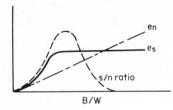

FIGURE 4-36 Effect of optimum bandwidth on signal to noise performance of an amplifier.

Even the "ideal" noiseless amplifier must always be coupled to a resistive source, which, at an effective noise temperature of 290 K, will thus offer a minimum detectable threshold of approximately −204 dBW/Hz. In the practical sense the amplifier will contribute noise which is, among other things, proportional to its bandwidth, resulting in the appearance at the output of an apparent threshold noise level whose magnitude is far greater than the ideal noiseless amplifier and perhaps even greater than the signal!

Previously we pointed out the importance of noise-matching the amplifier to offer optimum performance. However, an optimally matched input, by definition, forced narrow-bandwidth performance. Consequently, it appears that optimum bandwidth/optimum noise is a tradeoff design problem. Figure 4-36 represents how bandwidth affects signal/noise performance of the amplifier.

REFERENCES

1. Carson, Ralph S., "High Frequency Amplifiers," Wiley Interscience, New York, 1975.

2. Valley, George, and Henry Wallman, "Vacuum Tube Amplifiers," Radiation Laboratory Series, reprint by Dover Publications, New York, 1965.

BIBLIOGRAPHY

Ghausi, M. S.: *Principles and Design of Linear Active Circuits,* McGraw-Hill, New York, 1965.

Hetterscheid, W. Th. H.: *Transistor Bandpass Amplifiers,* Philips Technical Library, Eindhoven, the Netherlands, 1964.

Linvill and Gibbons: *Transistors and Active Circuits,* McGraw-Hill, New York, 1961.

Motorola Semiconductor Products Division, Phoenix, Ariz.:
Motorola Appl. Notes AN-423: "FET RF Amplifier Design Techniques."

Motorola Appl. Notes AN-215: "RF Small Signal Design Using 2-Port Parameters."

Motorola Appl. Notes AN-166: "Using Linvill Techniques for RF Amplifiers."

Raab, F. H.: "The Class BD High-Efficiency RF Power Amplifier," *IEEE J. Solid-State Circuits,* SC-**12**(3), June 1977.

————: "High Efficiency RF Power Amplifiers," *Ham Radio,* **7,** October 1974.

5

ANALOG SWITCHES

5-1   The FET as an Analog Switch
5-2   DC Equivalent Circuits
5-3   The JFET as a Switch
5-4   Switching High-Frequency Signals
5-5   The MOSFET Switch
5-6   The CMOS Switch
5-7   The VMOS Switch
5-8   DC Leakage Characteristics
5-9   Capacitance and Switching Transients
5-10   Signal Conversion Using Analog Switches
5-11   Conclusion

5-1 THE FET AS AN ANALOG SWITCH

The field-effect transistor, in the ON condition, contains a conducting channel of either n-type or p-type carriers. The carriers, in traversing from source to drain or drain to source, do not cross p-n junctions of the type encountered in bipolar transistor or diode switches; thus there is no inherent offset voltage in the ON switch. Signal currents can typically pass equally well in either direction in the ON switch and are blocked equally well in either direction by the OFF switch. This chapter presents some of the important characteristics of the FET as an analog switch and shows methods of driving the switch (controlling the ON-OFF status).

5-2 DC EQUIVALENT CIRCUITS

Some of the basic switching modes to be considered are shown in Fig. 5-1. The "voltage-mode" switch (a) and the "current-mode" switch (b), for example, may be used for multiplexing many signals into a common

193

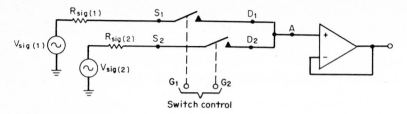

(a) Switch with high-impedance load

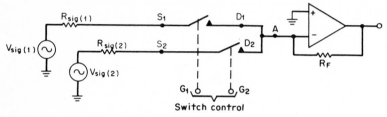

(b) Switch with low-impedance load

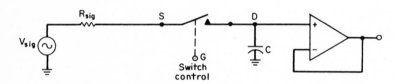

(c) Switch used in sample and hold system

FIGURE 5-1 Analog switch circuits: *(a)* switch with high-impedance load; *(b)* switch with low-impedance load; and *(c)* switch used in sample-and-hold system.

point such as the amplifier inputs shown. The Fig. 5-1*b* circuit is often used for the selective summing of two or more signals into a low-impedance summing node and in digital-to-analog converters. The Fig. 5-1*c*-type sample-and-hold circuit finds applications in analog-to-digital converters. For these circuits we will show how various FET types of the "family tree" may be used. We make the initial assumption that in the ON state the value of V_{DS} is small and may be of either polarity. For the FET types considered, the value of channel conductance is approximately a linear function of gate-to-source voltage, using as a reference the gate-source cutoff $V_{GS(off)}$ or threshold voltage $V_{GS(th)}$. Figure 5-2 shows this characteristic. Also shown in Fig. 5-2 is the effect that the body-source voltage has upon channel conductance of the MOS devices. The body, in effect, functions as a "back" gate, and its effect upon channel conductance must be considered.

The perfect switch would have infinite resistance (zero conductance) when open and zero resistance (infinite conductance) when closed. While the FET is not a perfect switch, there are many applications in which this deviation from perfection is unimportant. This statement can be justified by an analysis of the implications of the circuits shown in Fig. 5-3. The general two-port network in Fig. 5-3a couples the signal source V_{SIG} to a resistive load R_L. The network can be characterized by its terminal voltages and currents, V_1, V_2, I_1, and I_2. Figure 5-3b shows the equivalent circuit of a FET switch in the OFF state. In this condition, the "source" and "drain" are not connected to one another; however, two leakage current sources, I_S and I_D, are present. The same device is shown in the ON state in Fig. 5-3c. The FET gate is connected to its source. The leakage I_o is that of the driver. The following typical values are assumed for the circuit. The switch characteristics are those of the widely used analog switch type DG181AP.

$$V_{\text{SIG}} = \pm\,10\ \text{V} \qquad r_{DS} = 30\ \Omega$$
$$R_{\text{SIG}} = 10\ \Omega \qquad I_S = I_D = 1\ \text{nA}$$
$$R_L = 200\ \text{k}\Omega \qquad I_o = 2\ \text{nA}$$

In the following calculations the effect of leakage current is expressed in terms of error percentage.

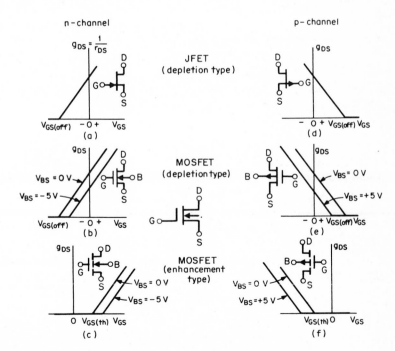

FIGURE 5-2   Channel conductance vs. gate-source voltage.

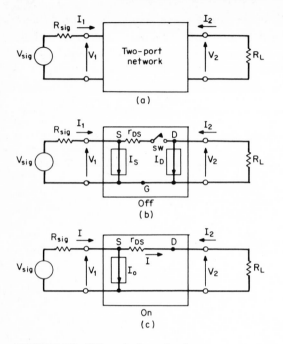

FIGURE 5-3   DC equivalent circuits.

5-2-1   OFF Condition Calculation (Fig. 5-3*b*)

$$I_1 = I_S = 1 \text{ nA}$$

$$V_{\text{SIG}} - V_1 = I_1 \cdot R_{\text{SIG}} = (1 \text{ nA})(10 \ \Omega) = 10 \text{ nV} \tag{5-1}$$

$$\% \text{ error in } V_1 = \frac{(10^{-8} \text{ V})(10^2)}{10 \text{ V}} = 1 \times 10^{-7}\%$$

$$I_2 = I_D = 1 \text{ nA}$$

$$V_{2(\text{off})} = I_2 R_L = (1 \text{ nA})(200 \text{ k}\Omega) = -200 \ \mu\text{V} \tag{5-2}$$

$$\% \text{ error in } V_{2(\text{off})} = \frac{(2 \times 10^{-4})(10^2)}{10} = 0.002\%$$

referred to V_{SIG} (full scale).

5-2-2   ON Condition Calculation (Fig. 5-3*c*)

$$I_2 = \frac{V_2}{R_L} \cong \frac{V_{\text{SIG}}}{R_L + R_{\text{SIG}} + r_{DS}} = 42 \ \mu\text{A} \tag{5-3}$$

$$V_{\text{SIG}} - V_2 \cong (42 \ \mu\text{A})(40 \ \Omega) = 1.7 \text{ mV}$$

$$\% \text{ error in } V_2 = \frac{(1.7 \times 10^{-3})(10^2)}{10} = 1.7 \times 10^{-2} = 0.017\%$$

referred to V_{SIG} (full scale)

The error resulting from I_o is approximately 20 nV or $2 \times 10^{-7}\%$.

The foregoing calculations indicate that for many applications the performance of the FET equivalent circuits in Fig. 5-3 is a good approximation of the perfect switch. In particular, the OFF condition leakage currents contribute only a negligible portion of total error.

5-3 THE JFET AS A SWITCH

A suitable driving circuit must be considered when assessing the performance of the JFET as a switch. Such a circuit is shown in Fig. 5-4.

Note that Q_1 is an n-channel JFET, Q_2 is an enhancement-mode p-channel MOSFET, and Q_3 is an enhancement-mode n-channel MOSFET. A V_{in} of -15 V will turn Q_3 OFF and Q_2 ON, so that S_1 and G_1 are connected. With $V_{\text{G1S1}} = 0$, Q_1 is ON. If V_{GS} of Q_1 were allowed to vary, switch resistance modulation would occur, introducing a source of error. Figure 5-5 shows the equivalent circuit of the ON switch.

The suggested driving circuit of Fig. 5-4 avoids r_{DS} modulation at low frequencies. Typically the positive supply voltage is $+15$ V and the negative supply voltage is -15 V. In order for V_{GS} to change, current must flow through Q_2, which is ON. There are only two possible current paths through Q_2: through Q_3, which is OFF and subject only to variations in leakage current, or into the gate of Q_2, which is also subject to leakage current. Since both paths through Q_2 provide only negligible changes in V_{GS}, their effect in the circuit may be ignored. As the analog frequency

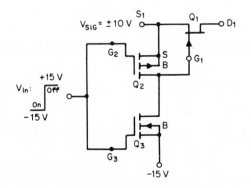

FIGURE 5-4 JFET switch control circuit.

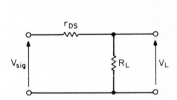

FIGURE 5-5 Error due to switch ON resistance r_{DS}.

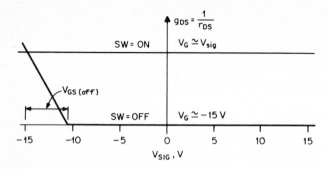

FIGURE 5-6 JFET channel conductance g_{ds} versus signal voltage.

is increased, capacitance will provide lower-impedance paths, so that some degree of Δr_{DS} is possible at high frequency.

With V_{in} at $+15$ V, Q_2 is OFF and Q_3 is ON; thus G_1 is at -15 V. Q_1 will remain OFF if $V_{SIG} > (V_{G1} - V_{GS(off)})$. $V_{GS(off)}$ is a "negative" voltage for an n-channel FET; thus the negative analog signal is limited by the $V_{GS(off)}$ of Q_1 and the negative supply voltage. The ON and OFF conditions are shown in Fig. 5-6. g_{DS} is constant because with $V_{G1} = V_{SIG}$ imposed by the switch control circuit, $V_{GS} = 0$.

The positive swing of the ON switch is limited only by the maximum voltage ratings of Q_2, and Q_3. In the OFF state the positive swing of the source and drain of Q_1 is limited by the voltage breakdown ratings of both Q_1 and Q_2.

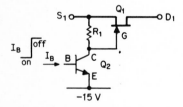

FIGURE 5-7 JFET switch control circuit.

Another driver circuit for the JFET is shown in Fig. 5-7. In the switch ON state, Q_2 is turned OFF; thus the gate of Q_1 is free to follow the analog signal at its source, *provided* the reactance of the gate-to-ground capacitance of Q_1 plus the output capacitance of Q_2 is large compared to the resistance of R_1. At high analog signal frequencies this circuit may develop problems because of the gate's inability to follow the source,

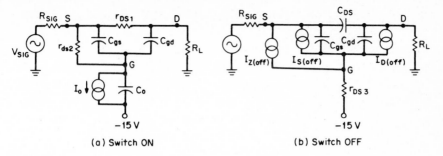

(a) Switch ON   (b) Switch OFF

FIGURE 5-8   Equivalent circuits using switch shown in Fig. 5-4.

which results in a modulation of the channel resistance by the signal voltage. This becomes more serious as the analog amplitude is increased.

In the switch OFF state, the gate of Q_1 is clamped to -15 V, and R_1 appears as a load on the analog signal source.

The ON and OFF equivalent circuits for the circuits of Figs. 5-4 and 5-7 are shown in Figs. 5-8 and 5-9. To minimize analog signal feedthrough of the OFF switch, r_{DS3}, r_{ce2}, C_{ds}, C_{gs}, and C_{gd} should be low. In the ON state, the value of r_{DS1} should be low to minimize insertion loss. Minimizing r_{DS1} must be compromised with keeping C_{gs} and C_{gd} low, because designing the FET for lower r_{DS} will usually increase its gate capacitance or increase $V_{GS(off)}$.

5-4   SWITCHING HIGH-FREQUENCY SIGNALS

As a signal frequency is increased, a decrease in OFF isolation rather than degradation of ON performance may become a limiting factor. Three factors affect the quality of OFF isolation: selection of the appropriate analog switch, the magnitude of load resistance, and the amount of stray capacitance present in the circuit. The selection of the appropriate

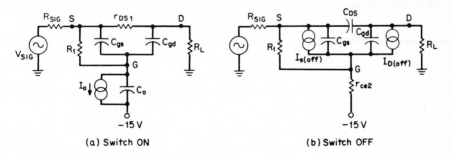

(a) Switch ON   (b) Switch OFF

FIGURE 5-9   Equivalent circuits using switch shown in Fig. 5-7.

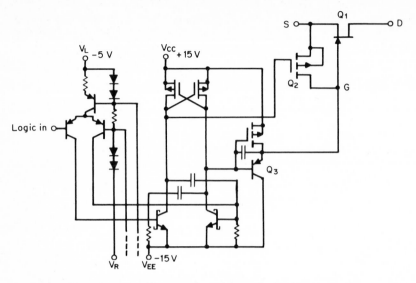

FIGURE 5-10 Schematic diagram of one channel of DG181BA analog switch.

analog switch is probably the most important factor. As a rule, product specifications do not provide OFF isolation data in sufficient depth to make switch selection easy. Three steps may be taken to overcome this lack of data: measurement of actual isolation performance, analysis of equivalent circuits to predict isolation, and simplification of the analysis to produce a usable set of design aids.

To illustrate performance analysis, a JFET switch with an integrated drive will be used for switching wideband RF signals. The JFET used as the switch is similar to the 2N3971 (except that $V_{GS(off)}$ is lower).

It is packaged with an integrated circuit driver which permits control with low-level logic signals such as the output from TTL digital circuits. A combined switch pair with driver has the commercial part number DG181BA. Figure 5-10 shows a schematic diagram of one channel of this dual-channel switch. The driver output is similar to the control circuit shown in Fig. 5-4. Replacing the MOSFET Q_3 of Fig. 5-4 with the *p-n-p* transistor shown in Fig. 5-10 improves OFF isolation because of the lower ON impedance of the *p-n-p* transistor.

5-4-1 OFF Isolation

The isolation performance test setup and a plot of data are shown in Fig. 5-11. In the equivalent OFF circuit shown in Fig. 5-12, C_o and R_o are output parameters of the driver *p-n-p* transistor Q_3 (Fig. 5-10). The 4.5-pF source-to-gate capacitance includes the source-to-drain capacitance of Q_2. The improvement in performance that could be obtained

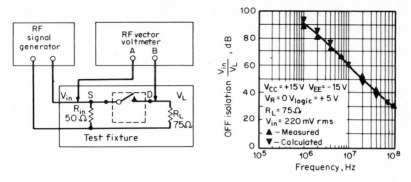

FIGURE 5-11   DG181 OFF isolation test circuit and data.

if R_o were reduced to zero is shown in the table included in Fig. 5-12. A large part of the feedthrough is via C_{gs} and C_{gd} unless the driver output impedance is low.

In Fig. 5-12, two separate paths exist between source *(S)* and drain *(D):* through C_{ds} and through the gate circuit C_{gs} and C_{gd} on the way to the drain. To simplify analysis, it is assumed that I_F (current through C_{ds}) and I_T (the current flowing out of the tee network) are independent

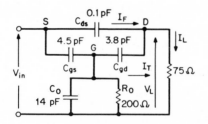

In the test fixtures, inputs were shielded from outputs, and RF decoupling was provided on all dc connections. Great care was taken in the mechanical layout of fixtures, to minimize stray capacitance. The characteristic impedance of video transmission lines, 75 Ω, was issued as the value of load resistance. Voltage measurements were made with an RF vector voltmeter, H/P Model 8405A.

COMPARISON OF IDEAL DRIVER CASE ($I_T = 0$) WITH ACTUAL PERFORMANCE OF DG181 ANALOG SWITCH

R_O	I_F	I_T	I_L	$\|V_{in}/V_L\|$ (dB)
0	1.41 $\underline{/90°}\,\mu$A	0	1.41 $\underline{/90°}\,\mu$A	66.5
200	1.41 $\underline{/90°}\,\mu$A	2.91 $\underline{/163°}\,\mu$A	3.58 $\underline{/141°}\,\mu$A	58.4

FIGURE 5-12   Equivalent circuit of DG181BA with 75-Ω load.

of one another. While this assumption is not entirely valid, if OFF isolation is greater than 20 dB, it yields excellent agreement with measured results.

The transfer functions for I_F and I_T are

$$I_F = \frac{j\omega C_{DS} R_L}{1 + j\omega C_{DS} R_L} \tag{5-4}$$

and

$$I_T = \frac{j\omega C_{GS} V_{in}}{1 + R_L/R_O + (C_{GS} + C_O)/C_{GD}} \\ {+ jR_L/R_O \left[\omega R_O(C_{GS} + C_O - 1/\omega R_L C_{GD}\right]} \tag{5-5}$$

R_O and C_O are the output resistance and capacitance of the FET driver circuit. Equation (5-5) shows that as $R_O \to 0$, $I_T \to 0$. It is thus possible to reduce R_O and make I_T an arbitrarily small value; however, I_F remains to be dealt with. I_F is the sum of the currents through C_{DS} (device capacitance) and C_{stray} (additional wiring capacitance, etc.). It may be the dominant current at certain frequencies. Table 5-1 shows that for the DG181 switch I_F is dominant at 1 MHz and I_T is dominant at 100 MHz. The separate expressions derived for I_F and I_T make it relatively simple to evaluate the effect of varying certain parameters to minimize I_L (maximize isolation).

Several multiple-switch configurations may be used to achieve an impressive increase in OFF isolation under otherwise difficult conditions. Probably the most effective multiple-switch configuration is the TEE, which is shown in Fig. 5-13. In the TEE, S_2 operates out of phase with S_1 and S_3, and provides two stages of isolation. The input to S_3 is the isolation leakage of S_1 working into an $R_L = r_{DS2}$. This multiple-switch arrangement brings about a considerable improvement in OFF isolation, but at the expense of doubling switch ON resistance. This increases the ON insertion loss.

TABLE 5-1 VARIATIONS IN CURRENT WITH FREQUENCY FOR DG181

f (MHz)	I_F	I_T	I_L	$\lvert V_{in}/V_L \rvert$ (dB)	$\lvert C_{eq} \rvert$ (pF)
1.0	141 $\underline{/90°}$ nA	30.3 $\underline{/178°}$ nA	145 $\underline{/102°}$ nA	+86	0.103
4.0	563 $\underline{/90°}$ nA	481 $\underline{/173°}$ nA	784 $\underline{/128°}$ nA	+72	0.139
10.0	1.41 $\underline{/90°}$ μA	2.91 $\underline{/163°}$ μA	3.58 $\underline{/141°}$ μA	+58	0.254
40.0	5.63 $\underline{/90°}$ μA	31.9 $\underline{/128°}$ μA	36.5 $\underline{/123°}$ μA	+38	0.648
100.0	14.1 $\underline{/90°}$ μA	99.6 $\underline{/101°}$ μA	113 $\underline{/100°}$ μA	+28	0.803

$V_L = 224$ mW; $R_L = 75$ Ω; $C_{DS} = 0.1$ pF; $C_{stray} = 0$.

NOTE: The equivalent circuit shown in Fig. 5-12 was used to calculate the results shown in Table 5-1.

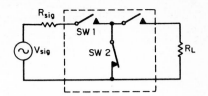

FIGURE 5-13 TEE for isolation improvement.

5-4-2 ON Attenuation

The ON equivalent circuit of a DG181 switch is shown in Fig. 5-14. ON performance is essentially independent of frequency for any load capacitance likely to be used. The ON resistance $r_{DS(on)}$ causes an insertion loss which is basically constant; phase shift is negligible.

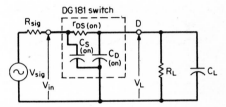

FIGURE 5-14 DG181 ON equivalent circuit.

With the test fixture described in Fig. 5-11, measured ON performance was observed as shown in Fig. 5-15.

The transfer function for the ON switch is

$$\frac{V_L}{V_{in}} = \frac{R_L/(R_L + r_{DS(on)})}{1 + jf\{2\pi[(R_L r_{DS(on)})/(R_L + r_{DS(on)})][C_{D(on)} + C_L]\}} \qquad (5\text{-}6)$$

$$f_0 = \frac{1}{2\pi\left(\dfrac{R_L\, r_{DS(on)}}{R_L + r_{DS(on)}}\right)\left(C_{D(on)} + C_L\right)}$$

FIGURE 5-15 DG181 ON performance.

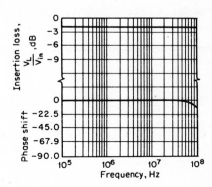

Insertion loss is computed from the numerator as (neglecting C_D and C_L)

$$\text{ON insertion loss, dB} = 20 \log \frac{R_L}{R_L + r_{DS(on)}} \tag{5-7}$$

The insertion loss, OFF isolation, and ON/OFF ratio over a range of load resistance are

Switch	$r_{DS(on)}$ (Ω)	R_L (Ω)	Insertion loss (dB)	OFF isolation (dB) $f = 10\ MHz$ $C_{stray} = 0.1\ pF$	ON/OFF (dB)
DG181	25	100	2.0	55.9	53.9
		75	2.5	58.4	55.9
		50	3.6	61.9	58.3

The frequency response of the ON transfer function has the normalized form shown in Fig. 5-16.

5-5 THE MOSFET SWITCH

The control circuit for the MOSFET switch can be simpler than for the JFET because its gate may be positive or negative with respect to source and drain. Typically the gate is switched between two fixed voltages, the sum of which must exceed the peak-to-peak analog voltage by at least the magnitude of $V_{GS(off)}$ or $V_{GS(th)}$ plus enough additional

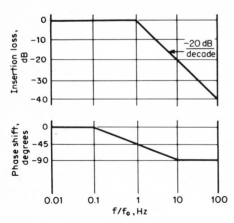

FIGURE 5-16 ON frequency response.

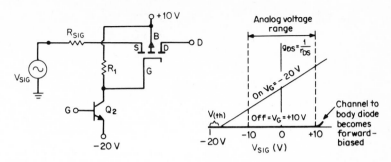

FIGURE 5-17   PMOS channel conductance g_{ds} versus signal voltage.

V_{GS} bias to produce an adequate g_{ds}. As an example, consider the p-channel enhancement-mode MOSFET switch and driver shown in Fig. 5-17. The gate is connected to +10 V (through R_1) for the switch OFF state and to −20 V (via Q_2) for the switch ON state. In the ON state, the channel conduction g_{ds} is a function of the signal voltage as indicated because both gate and body are connected to fixed voltages (−20 V and +10 V). If the source or drain is allowed to swing more positive than +10 V, they become forward-biased with respect to the body, a condition which should typically be avoided. If source and drain approach −20 V, channel conductance g_{ds} will approach zero.

5-6   THE CMOS SWITCH

As noted in Fig. 5-17, the typical PMOS or NMOS switch circuit will exhibit a variation in ON conductance as the analog voltage is varied. This undesirable characteristic can be overcome by paralleling p- and n-channel MOSFETs, as shown in Fig. 5-18a. For the ON state, the n-channel gate is forced positive and the p-channel gate is forced negative. Figure 5-18b shows the combined conductance of the two FET switches. The integrated combination of n-channel and p-channel devices on a common substrate is referred to as complementary MOS (CMOS).

The OFF condition for the CMOS device will be maintained so long as the channel-to-body diodes do not become forward-biased, as shown in Fig. 5-18c.

The major advantages the CMOS construction technique makes to analog switching are:

○ Lower r_{DS} variation than with either P or NMOS over wide analog signal voltage excursions (similar to the performance of a junction FET).

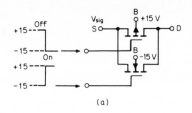

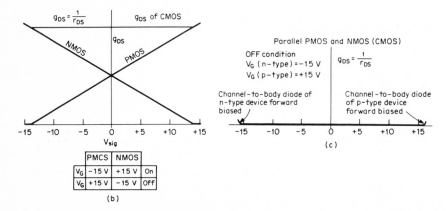

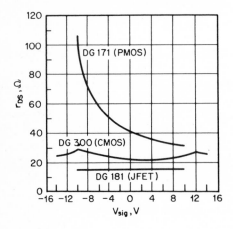

FIGURE 5-18   Characteristics of CMOS devices.

○ Analog signal range extends to + and − supply voltages. For instance, using the same ±15-V supplies typical of operational amplifiers, the signal-handling capability of the system is limited by the op amp, *not by the switch.*

Figure 5-19 gives a comparison of the characteristic of r_{DS} versus V_{SIG} for typical JFET, PMOS, and CMOS switches. The three characteristic

FIGURE 5-19   Performance of three FET switches.

curves shown in Fig. 5-19 were obtained using integrated-circuit analog switches types DG181, DG171, and DG300, respectively, for the JFET, PMOS, and CMOS types.

5-7 THE VMOS SWITCH

The low ON resistance of the VMOS when used as an analog switch results in low insertion loss in low-impedance systems, fast charging in sample-and-hold systems, rapid discharge of integrator capacitors, low noise in measuring systems, and high accuracy in test systems. Its high current-carrying capability allows transmission of considerable power through the switch, and ease of paralleling without ballast resistors increases this capability. The ability to carry high peak currents is advantageous for driving capacitive lines and quickly charging and discharging capacitors in high-speed A/D converters, S/H circuits, and integrators. Their high OFF isolation (greater than 60 dB isolation at 10 MHz) and less than 500 nA dc leakage provide excellent OFF characteristics. VMOS are fast—a characteristic useful in radar, sonar, and laser applications, where signal bursts must be gated ON and OFF very rapidly. A typical VMOS can switch 1 A in 4 ns. Their linear ON resistance results in low total harmonic and intermodulation distortion. This is especially important in color video switching, where color purity must be maintained.

Figure 5-20 shows the r_{DS} characteristics of a low-resistance VMOS device, the 2N6659. Varying the gate voltage from zero to +10 V switches the 2N6659 from an OFF condition to less than 2 Ω ON resistance.

A schematic representation of the 2N6659 and its equivalent OFF and ON circuits is given in Fig. 5-21. It is important to note that the body of the device is internally connected to the source. Diode D_1 is the body-drain p-n junction.

The drain current versus drain-to-source voltage characteristic in the

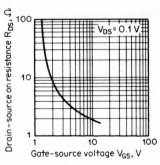

FIGURE 5-20 Drain-to-source ON resistance vs. gate-to-source voltage.

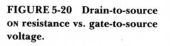

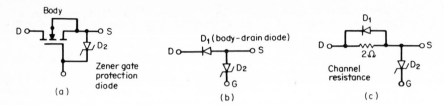

FIGURE 5-21 *(a)* **Schematic symbol of 2N6659;** *(b)* **equivalent OFF condition** *($V_{GS} =$* **0); and** *(c)* **equivalent ON condition** *($V_{GS} = +10$ V).*

OFF condition (Fig. 5-22) has the appearance of a diode characteristic. The drain current is very low until the drain-source is reverse-biased to about 0.6 V. In the ON condition the channel conductance of about 0.5 mho parallels the diode.

A practical implementation of this device as an analog switch is shown in Fig. 5-23. In the ON condition the gate of the 2N6659 is positive with respect to the source, whereas in the OFF condition the gate voltage is zero. This circuit can take advantage of the 1-A capability and 2-Ω ON resistance of the VMOS. However, in the OFF state the input signal is restricted to positive voltages and should always be greater than the output voltage; OFF isolation is impaired otherwise because of the drain to body diode. Switching times are less than 200 ns, and charge feedthrough during the ON to OFF transition is 80 pC with a 50-Ω load. Charge transfer is important in sample-and-hold systems, where an offset voltage of 8 mV into a 0.01-μF load would occur in this case.

To increase the dynamic range to ± 10 V, two 2N6659s are connected in series source-to-source as shown in Fig. 5-24. In the ON condition, both output switches of the DG300 are open, and the gates of both

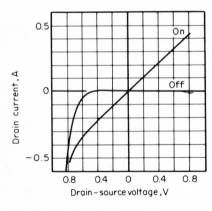

FIGURE 5-22 Small-signal characteristics of 2N6659.

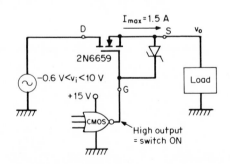

FIGURE 5-23 A simple unidirectional VMOS analog switch *($V_i \geqslant V$).*

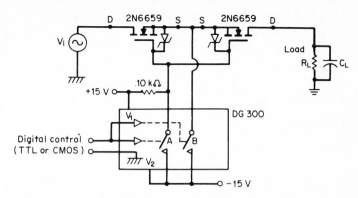

FIGURE 5-24 A general-purpose bidirectional analog switch.

2N6659s are pulled to +15 V. The ON resistance of the switch is now twice the drain-source resistance of a single 2N6659, but the maximum current is still that of a single device. The switch is turned OFF by shorting the VMOS gates to the negative supply, reducing V_{GS} to a voltage less than the minimum threshold of 0.8 V. Switch B of the DG300 increases OFF isolation 30 dB by shunting the signal leakage path (through the sources of the 2N6659s) to the negative supply.

OFF isolation is shown in Fig. 5-25. The previous problem with the body drain diodes forward-biasing in the OFF condition is now removed,

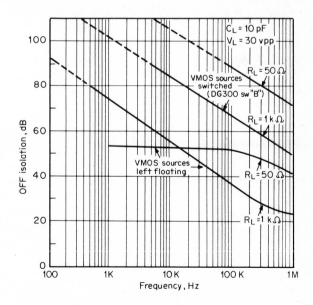

FIGURE 5-25 OFF isolation vs. frequency.

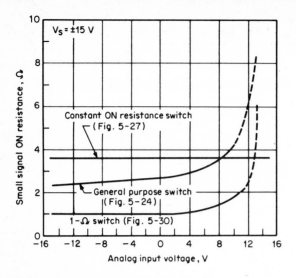

FIGURE 5-26 Small-signal ON resistance vs. analog
input voltage.

since the two FETs are back-to-back, so one diode is always reverse-
biased.

The bidirectional switch has a gate drive which is referenced to a
fixed supply. Its ON resistance varies with the input analog voltage because V_{GS} changes (Fig. 5-26). This variation may introduce distortion
when driving low-impedance loads such as speakers or transmission lines.
To achieve a constant ON resistance, use the circuit shown in Fig.
5-27. In the ON condition, the gates of the 2N6659s are driven by a

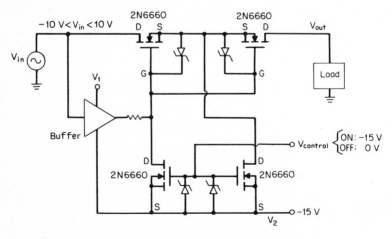

FIGURE 5-27 Low-distortion constant-ON-resistance switch.

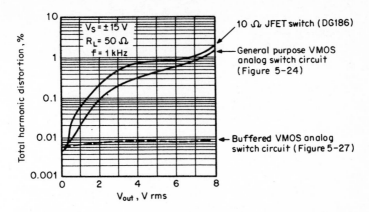

FIGURE 5-28 Distortion improvement using the buffered analog switch.

voltage which tracks the input signal, so gate-to-source voltage is constant and independent of the input signal. No modulation of the ON resistance therefore takes place as the signal level changes. The buffer circuit reduces the total harmonic distortion (THD) from 1.5 percent (no buffer) to less than 0.005 percent (calculated) at 1 kHz, 8 V rms into 50 Ω. The next best analog switch, a 10-Ω DG186, has a 2 percent THD (Fig. 5-28).

The two buffer circuits shown in Fig. 5-29 isolate the input signal

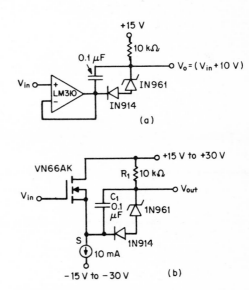

FIGURE 5-29 (a) **General-purpose buffer and** (b) **high-speed buffer.**

and use a zener diode to provide a fixed $V_{GS(on)}$ voltage. The general-purpose buffer is flat to 300 kHz and operates on ±15 V or less. The second buffer, a VN66AK voltage follower, shown in Fig. 5-29b, extends the frequency response to 50 MHz and increases the signal range to ±30 V when operated from ±30-V supplies. The use of a bootstrap in the buffer circuit allows large-signal, low-distortion operation near the positive supply, provided the switch-ON time is small compared to the time constant of $R_1 C_1$.

Lower ON resistance is easily achieved by paralleling devices. For example, three paralleled 2N6659s result in a 1-Ω switch (Fig. 5-30). No ballasting or balancing resistors are needed, since the 2N6659s do not suffer from current hogging. VMOS has a temperature coefficient which causes it to draw less current as it heats up, so excess current is automatically shifted to the other devices. Paralleling the 2N6659 not only decreases the ON resistance but increases the current-carrying capability to 4.5 A and the linear transfer characteristic to 1.2 A (Fig. 5-31). The voltage range can also be increased. Simply use higher breakdown VMOS as shown in Fig. 5-32. The VN98AK has a breakdown of 90 V, allowing up to ±40 V analog capability. However, its ON resistance is also greater (2.8 Ω vs. 1.2 Ω for the 2N6659). Zener D_1 limits the gate-source voltage to 30 V, preventing possible gate oxide rupture.

The high-power RF switch shown in Fig. 5-33 has excellent perform-

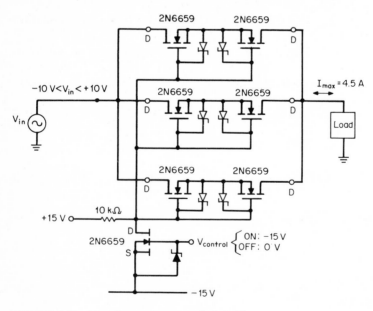

FIGURE 5-30 Ultra-low-resistance switch (1 Ω).

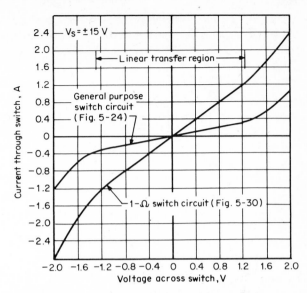

FIGURE 5-31   Large-signal transfer characteristics.

ance up to 50 MHz, with turn-ON and turn-OFF times of less than 50 ns. The isolation at 10 MHz is better than 60 dB with a 20 V p-p input signal; insertion loss is only 1 dB with a 50-Ω load (Figure 5-34a). The gain versus input power and the two-tone third-order intermodulation product performance curves (Fig. 5-34b) show up to 1 W of power being transferred to the 50-Ω load with a 42-dB intercept point and 1 dB gain compression at 25 dBm input power. The turn-ON time of the switch (Fig. 5-34c) is determined by the passive pullup resistor in combi-

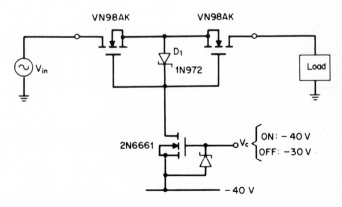

FIGURE 5-32   80-V peak-to-peak analog switch.

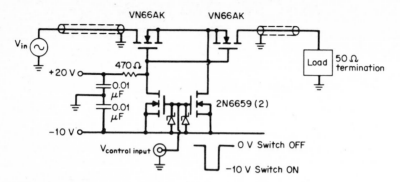

FIGURE 5-33   RF analog switch.

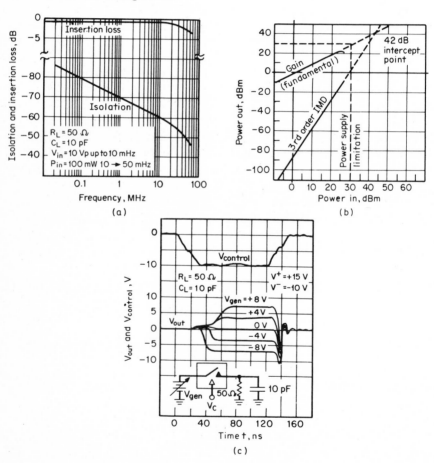

FIGURE 5-34 *(a)* Insertion loss and isolation vs. frequency of RF analog switch; *(b)* gain and two-tone, third-order intermodulation; and *(c)* switching response of RF switch into 50-Ω load.

nation with the capacitance at the gates of the VN66AK; the negative turn-OFF transient is caused by charge coupling to the output through the output capacitance C_{oss} of the 2N6659.

5-8 DC LEAKAGE CHARACTERISTICS

The ON and OFF dc leakage of the analog switch circuit is a function of the FET switch plus its driver circuit. The calculation of system error due to leakage was presented earlier in dc equivalent circuits. For the JFET switch, the sources of the error-producing leakage currents are shown in Figs. 5-8 and 5-9. In the ON condition, the switch gate is clamped to its source and follows the analog signal. Signal leakage is principally to the −15-V supply due to the OFF leakage of the driver (Q_3 of Fig. 5-4 or Q_2 of Fig. 5-7). It is shown as I_o in Figs. 5-8a and 5-9a. If the signal source resistance R_{SIG} is low, then usually the effect of I_o and $I_{S(off)}$ may be neglected, typical values being less than 10^{-9} A at 25°C. With a value of R_{SIG} of 100 Ω, this magnitude of leakage would result in an error of 100×10^{-9} V.

$I_{D(off)}$ may be a problem in multiplexers where many switches are connected to a common output, such as point A in Figs. 5-1a or b. Then the sum of the switch leakages must be considered. For example, examine the multiplexer circuit of Fig. 5-35a. Assume that one switch is turned ON, $I_{D(off)}$ is the same for all switches, I_o is the leakage of the ON switch driver, and the input conductance of the op amp is zero.

For this case, the error voltage due to leakage current is

$$V_{A(error)} = -[(n-1) I_{D(off)}](r_{DS(on)} + R_{SIG}) - I_o R_{SIG} \qquad (5\text{-}8)$$

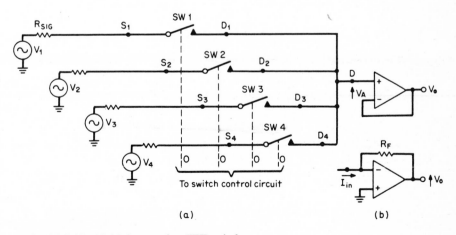

(a)   (b)

FIGURE 5-35 Multiplexer using FET switches.

If, for example, $n = 16$, $r_{DS(on)} = 30\ \Omega$, $R_{SIG} = 100\ \Omega$, $I_{D(off)} = 0.25$ nA, and $I_0 = 1$ nA, the error voltage will be

$$V_{A(error)} = -[(15)(0.25)(10^{-9})(30 + 100)] - (10^{-9})(100) \cong -0.59\ \mu V$$

For high-temperature operation, say at 125°C, $I_{D(off)}$ may increase to 25 nA, $r_{DS(on)}$ may increase to 55 Ω, and I_0 to 100 nA. The error voltage then would increase to approximately $-68\ \mu V$.

Equation (5-8) shows that as the number of switch channels in a multiplexer is increased, the error due to $I_{D(off)}$ becomes more significant. By assuming that $I_0 = I_{D(off)}$, $n \gg 1$, and $R_{SIG} \cong r_{DS(on)}$, (Eq. 5-8) can be simplified:

$$V_{A(error)} \cong nI_{D(off)}(R_{SIG} + r_{DS(on)}) \qquad (5\text{-}9)$$

If the output of the multiplexer is connected to a current-summing node, such as the inverting amplifier shown in Figs. 5-35b and 5-36, then the error voltage is due to leakage currents into this low-impedance node. Assuming that $R_{SIG} \gg r_{DS(on)}$, which is typically true for this type of application, then

$$V_{o(error)} \cong [I_0 + (n-1)I_{D(off)}]R_F \qquad (5\text{-}10)$$

when $V_{o(error)}$ is the output voltage resulting from switch leakage currents.

5-8-1 Two-Level Multiplexer

If the number of channels n is increased to the extent that the error due to leakage current is excessive, then a two-level multiplexing system may be used. Figure 5-37 shows a two-level system with its equivalent

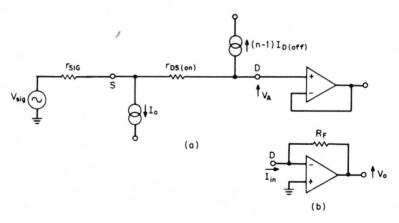

FIGURE 5-36 Equivalent circuit of multiplexer with one switch ON and $n - 1$ switches OFF.

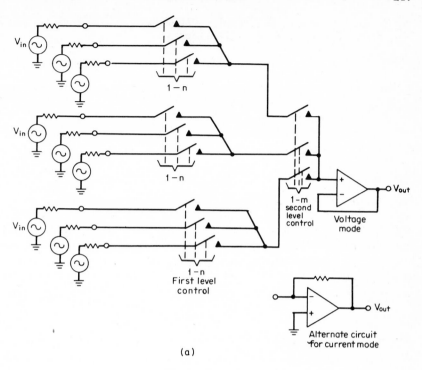

(a)

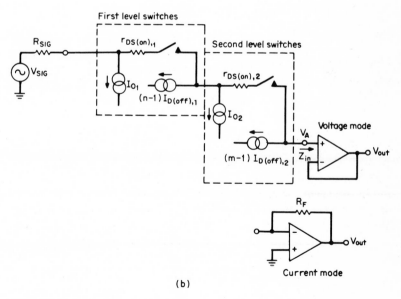

(b)

FIGURE 5-37 *(a)* Two-level multiplexer system and *(b)* equivalent leakage circuit of two-level multiplexer.

circuit used to determine error due to leakage. For the "voltage-mode" switch, using a high-Z_{in} op amp load, the approximate error voltage due to leakage is

$$V_{A\,(error)} \cong -[(m-1)\,I_{D\,(off)_2}(r_{DS1} + r_{DS2} + R_{SIG})]$$
$$- [I_{o2} + (n-1)I_{D\,(off)}](r_{DS1} + R_{SIG}) - I_{o1}R_{SIG} \quad (5\text{-}11)$$

This assumes that $Z_{in} \gg (r_{DS1} + r_{DS2} + R_{SIG})$.

For the current-mode switch, using a low-Z_{in} inverting amplifier, the error due to leakage is

$$V_{o\,(error)} = [I_{o1} + (n-1)\,I_{D\,(off)_2} + I_{o2} + (m-1)\,I_{D\,(off)_2}]\,R_F \quad (5\text{-}12)$$

By way of example, compare a 1024-channel single-level system with a 1024-channel two-level system in which $n = m = 32$. Assume that at the maximum operating temperature $I_{D\,(off)} = I_o = 50$ nA, $r_{DS\,(on)} = 50\ \Omega$, and $R_{SIG} = 100\ \Omega$. For the single-level system, using Eq. (5-9),

$$V_{A\,(error)} \cong 7.68\ \text{mV}$$

For the two-level system, using Eq. (5-11),

$$V_{A\,(error)} \cong 0.55\ \text{mV}$$

The two-level system has the added advantages of reducing the number of control lines from $n \times m$ lines to $n + m$ lines and reducing capacitance at the output node by approximately the same factor by which leakage current effects were reduced.

5-8-2   MOSFET Leakage

The gate leakage of the MOSFET in a typical analog switching circuit may be neglected when compared to the drain-to-body and source-to-body leakage; thus the driver circuit does not contribute directly to leakage errors. If MOS switches are used in the multiplexer of Fig. 5-35, the equivalent circuit of Fig. 5-38 will be helpful in analyzing the leakage errors.

In the normal operating mode ($V_{SIG} \leq V_U$) the leakage currents are due to the reverse-biased source-body and drain-body diodes. The leakage of the ON switch is only slightly greater than $I_{D\,(off)} + I_{S\,(off)}$. Also, for the typical switch, $I_{D\,(off)} \cong I_{S\,(off)}$. By inspection of Fig. 5-37 (assuming one switch is ON), the error voltage due to leakage is

$$-V_{A\,(error)} = nI_{D\,(off)}(r_{DS\,(on)} + R_{SIG}) + I_{S\,(off)}R_{SIG} \quad (5\text{-}13)$$

A factor of the MOS switch that may not be obvious is its characteristics in the event of an overvoltage condition where $V_S > V_U$ or $V_D > V_U$. Due to the close proximity of the source to the drain, one can function

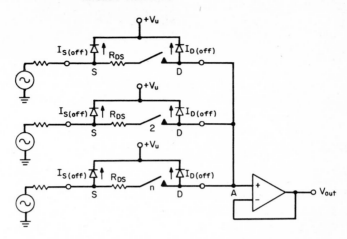

FIGURE 5-38 Equivalent circuit of multiplexer using MOSFET switches.

as an emitter and the other as a collector, with the "body" serving as a base to form a *p-n-p* junction transistor. This means that most of the current injected into the body by a forward-biased source-body junction may be "collected" by a reverse-biased drain. If the source-body or drain-body "diode" becomes forward-biased, the MOSFET no longer provides good OFF source-drain isolation. The analog voltage should not exceed the positive body supply voltage.

5-8-3 CMOS Switch Leakage

The CMOS switch shown in Fig. 5-18 has a *p*-channel MOS and an *n*-channel MOS in parallel. In its OFF state, the PMOS body is connected to the positive supply and the NMOS body to the negative supply. The results in the PMOS leakage being opposed by the NMOS leakage. The net leakage may be zero at a particular value of drain voltage. The voltage at which this occurs is usually very near the value of the positive supply because in a complementary switch the PMOS is larger than the NMOS and thus has a higher leakage.

5-9 CAPACITANCE AND SWITCHING TRANSIENTS

The capacitance associated with the steady-state ON and OFF states of the FET switch was discussed in the switching high frequency section as it related to signal attenuation and isolation. In this section, we will evaluate the transients which occur during the turn-ON and turn-OFF periods.

By the very nature of its operation the FET requires a certain transfer of charge to change its state from a conducting to a nonconducting mode. The required charge transfer is related to the channel thickness, length, and width and carrier concentration within the channel. Thus, it is related to $r_{DS(on)}$ and $V_{GS(off)}$. For switching applications, low values of $r_{DS(on)}$ and $V_{GS(off)}$ are desirable for charge transfer; however, there must be a design compromise between these conflicting parameters. Increasing channel width will reduce $r_{DS(on)}$ without affecting $V_{GS(off)}$; however, the charge required to turn the device OFF will be increased. Increasing channel thickness will also decrease $r_{DS(on)}$, but increase $V_{GS(off)}$ and thus charge transfer. A reduced channel length results in a decrease in both $r_{DS(on)}$ and charge transfer without an appreciable effect on $V_{GS(off)}$; thus good switching FETs usually have short channels.

5-9-1 Sample-and-Hold Circuit

The charge transfer characteristic is of particular importance in sample-and-hold-type circuits because any charge transferred into or out of the holding capacitor in the transition between the sample-and-hold mode results in an error signal. This characteristic will vary to quite an extent, depending upon various switch and circuit configurations. Few, if any, device data sheets include as a characteristic the charge in picocoulombs required to turn the device ON or OFF. A rough estimate can be had by multiplying $V_{GS(off)}$ and $C_{gs} + C_{gd}$. Caution must be used in comparing units on this basis because the values of C depend upon

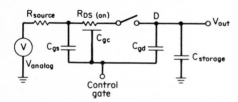

FIGURE 5-39 Equivalent switch circuit for a sample and hold.

bias conditions. Consider Fig. 5-39 as the equivalent circuit for a sample-and-hold circuit. In the switch ON state $C_{storage}$ will charge to the value of V_{analog}. Ideally when the switch is turned OFF, the storage capacitor will hold at whatever value V_{analog} had at the instant the switch was opened. In practice, however, a charge will be transferred from the

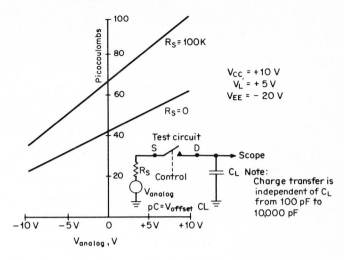

FIGURE 5-40 Typical charge transfer characteristic of the DG181 JFET switch.

control gate through the switch capacitance into the storage capacitance. This will result in an offset (error) voltage.

$$\text{Voltage offset} = \frac{\text{charge transfer (picocoulombs)}}{\text{hold capacitance (picofarads)}} = \frac{Q}{C} \quad (5\text{-}14)$$

Using the JFET switch with an integrated driver circuit that was used in the high-frequency example (see Fig. 5-11), we will examine the charge transfer as a function of the analog voltage level. Figure 5-40 shows this characteristic for the analog switch type DG181. In the OFF state the FET gate is clamped to approximately -20 V. In the ON state the gate is essentially at the analog voltage. In going from ON to OFF, then, the control gate has a voltage swing of $V_{EE} - V_{analog}$ volts. The curves of Fig. 5-40 show an increase in charge transfer as V_{analog} is increased. The charge transfer is lower for the case where R_S is zero because a part of the charge is shunted to ground through the signal source. This shows that a lower signal-source resistance results in a lower error. Equation (5-14) indicates that increasing the value of the holding capacitance will reduce the voltage offset resulting from charge transfer.

The JFET construction provides a good compromise between low ON resistance and low coupling capacitance. A comparison of a PMOS switch with the JFET characteristic will show a major difference. The charge transfer characteristic of a PMOS switch type DG172 is given in Fig. 5-41. The switch control circuit is similar to that shown in Fig. 5-18.

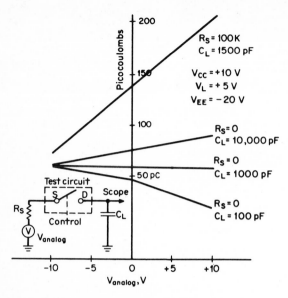

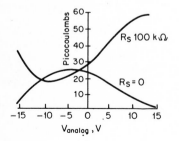

FIGURE 5-41   Typical charge transfer characteristics of the DG172 PMOS switch.

As may be seen from the two curves of Figs. 5-40 and 5-41, the charge transfer characteristics for the DG181 and DG172 are similar at $C_L = 100$ pF and $R_S = 0$. It should be noted, however, that the DG172 has a typical ON resistance of 200 Ω—an increase in ON resistance of six times that of the DG181. Values of capacitance greater than 100 pF in the charge transfer characteristics of the DG172 are seen to be inferior to those of the DG181. The major factor which causes the storage capacitance to be value-dependent is the large distributed gate-to-channel capacitance, plus the related circuit time constants.

CMOS devices provide an improvement over the JFET and PMOS devices since two gates with complementary control signals are involved. The two resulting charge transfers tend to cancel each other. The tran-

FIGURE 5-42   CMOS charge transfer characteristics.

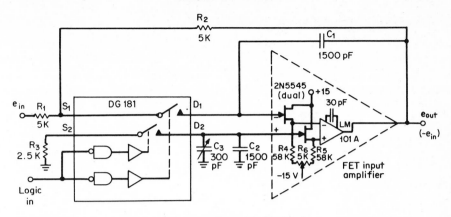

FIGURE 5-43 Improved inverting sample-and-hold circuit.

sient is therefore greatly reduced but not eliminated. This is shown in Fig. 5-42.

An example of a means of compensating for the charge transfer is shown in Fig. 5-43. In this circuit the charge transfer into the positive and negative inputs of the inverting amplifier is nearly equal; thus the net output voltage resulting from charge transfer is nearly zero. Capacitor C_3 provides a means of adjusting for the nonequality of the two switches in the DG181. (The particular layout of the DG181 results in a slightly higher charge transfer into D_2 than into D_1.)

An added feature of this circuit is that the balancing effect of the switch leakage currents results in a reduction in the voltage drop. For designs requiring faster settling times, the noninverting circuit shown in Fig. 5-44 may be used.

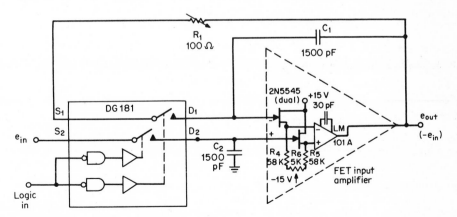

FIGURE 5-44 High-performance noninverting sample-and-hold circuit.

5-9-2 Switching Spikes

In switching circuits not involving capacitive loads, the charge transfer during turn-ON and turn-OFF results in a transient voltage or current "spike" or "glitch." The FET switch itself is a high-speed device; the turn-ON and turn-OFF time is essentially that required to accomplish the necessary gate-charge transfer. These times are typically determined by the driver circuit. The faster the switch is toggled, however, the greater will be the peak amplitude of the spike caused by the charge transfer.

The amplitude is also a function of the analog-signal source and load impedances. Typically for the "voltage-mode" switch (Fig. 5-13), R_{SIG} is small so that the transient at the input does not contribute significant error to the system. The apparent load impedance during the switching period will depend upon the status of the other switches in the system. For example, the output spikes will be lower if the switch of Fig. 5-13 is a "make-before-break" than if it is "break-before-make."

The equivalent circuit of two channels is shown in Fig. 5-45. Assuming the switch is an n-channel JFET with driver similar to that used in the DG181 (see Fig. 5-10), the turn-ON output spike will be positive and the turn-OFF output spike will be negative. Switch 1 of Fig. 5-45 has as an output load the parallel combination of R_L, C_L, and switch 2. If switch 2 is OFF, then its loading effect is essentially the capacitance C_{gd2}. Assuming R_{SIG} is low, the spike appearing at the output is due mainly to charge transfer from the driver through C_{gd1}. In turning switch 1 from ON to OFF, part of the charge will flow through $r_{DS(on)}$ and R_{SIG}

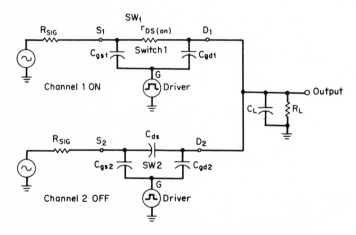

FIGURE 5-45   Equivalent circuit of two-channel multiplexer.

to ground, so that initially the effective load consists of R_L, C_L, and $(r_{DS} + R_{SIG})$ in parallel.

As the gate becomes more negative, r_{DS} will increase and approach infinity as switch 2 turns OFF; thus $(r_{DS(on)} + R_{SIG})$ will no longer shunt the output. At the same time, C_{gd} is decreasing because V_{DG} is increasing. These two variables plus the lack of a well-defined driver output pulse shape make detailed analysis of the output pulse difficult. Such an analysis has been presented in the analog switch handbook referenced above.

An approximation of the spike output amplitude can be made by utilizing the charge transfer curves shown in Figs. 5-40 and 5-41, the analog voltage level, and the output load. In the simplified equivalent circuit shown in Fig. 5-46, the effective value of C_{gd} is $Q/\Delta V_{GS}$, where Q is the charge transfer in coulombs, ΔV_{GS} is the change in V_{GS} ($V_{analog} - V_{EE}$), and C_{gd} is in farads. For example, let $V_A = 0$. From the curve of Fig. 5-41, $Q = 40$ pC. Since $V_{EE} = -20$ V, $\Delta V_{GS} = 20$ V and

$$C_{gd(off)} = \frac{40 \text{ pC}}{20 \text{ V}} = 2 \text{ pF}$$

If the R_{SIG} is not zero, then the $C_{gd(off)}$ will be increased because the charge lost through the analog source will decrease. From the $R_{SIG} = 100$ kΩ curve of Fig. 5-25, $Q \cong 66$ pC at $V_A = 0$ and

$$C_{gs(off)} = \frac{66 \text{ pC}}{20 \text{ V}} = 3.3 \text{ pF}$$

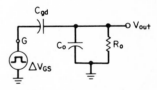

FIGURE 5-46 Simplified circuit for charge transfer analysis.

Figure 5-47 shows typical switching transients. Recovery or settling time after the turn-OFF transient is essentially the time constant of $R_o C_o$; however, in a multiplexer application, when one channel is turned OFF, another channel is turned ON, so that the settling time is a function of $r_{DS(on)}$ rather than R_o. Figure 5-48 shows a 32-channel two-level multiplexer with an output settling time of less than 200 ns.

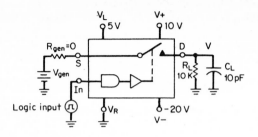

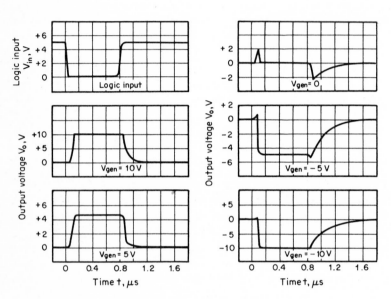

FIGURE 5-47 Typical switching times and transients.

5-10 SIGNAL CONVERSION USING ANALOG SWITCHES

Extensive use of the FET as an analog switch is made in signal conversion systems such as analog-to-digital (A/D) and digital-to-analog (D/A) converters. Increased use of digital techniques to replace analog systems has resulted in the evolution of a multiplicity of converter types. There are still occasions where nonstandard digital codes or analog voltages, a modification of an existing design, or a new design is required. In such cases, FET analog switches as described in this chapter may be useful. Examples are presented below.

5-10-1 Weighted Network Converter

The majority of D-to-A converters available today are based on the weighted network principle. An example of this in binary form is shown

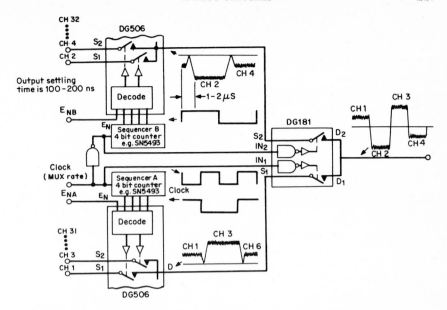

FIGURE 5-48 Practical two-level multiplexer.

in Fig. 5-49. Weighted currents are summed at the output node of the network. The currents are produced from a voltage reference source via an analog switch and a resistor. The value of the resistor depends on the "weighting" of the digit driving it. In the above example, the current produced when the most significant bit is energized is V_{REF}/R. The second most significant bit operation produces a current $V_{REF}/2\,R$, and the third most significant bit operation produces a current $V_{REF}/4\,R$.

In this case the currents are binary weighted. If the resistor values were in decade ratios, the currents would also be in decade ratios. The main drawback of this system is the large values of resistance that may be necessary. For instance, for a 10-bit binary-weighted network with $R = 1\ k\Omega$, the tenth bit has a value $1\ k\Omega \times 2^{10} = 1.024\ M\Omega$. To maintain

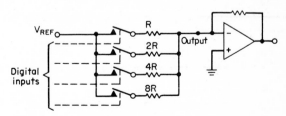

FIGURE 5-49 Binary-weighted resistor D/A converter.

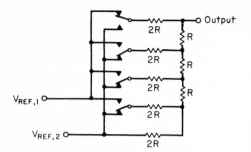

FIGURE 5-50   Ladder network D/A converter.

the accuracy of the conversion over a temperature range, the resistor ratios must track. This is difficult with resistor ratios of 1000 to 1. Also, the impedance of the network varies with the input code. This can upset the offset and drift of any following amplifier, and hence system accuracy.

5-10-2   Ladder Network Converter

A resistor system that achieves the same "weighting" performance without the large range of values is the binary-weighted ladder network with switches shown in Fig. 5-50. The resistor values have $2:1$ ratios. The operation of the ladder network can be understood by considering a simple two-bit converter with V_{REF2} equal to zero volts (shown in Fig. 5-51). With the most significant bit (MSB) equal to logic 1 (V_{REF1}) and the least significant bit (LSB) equal to logic "0" (V_{REF2}), the output current I_z will be $V_{REF1}/2R$.

In Fig. 5-52, with MSB = logic 0 and LSB = logic 1 (V_{REF1}), the output current I_z can be calculated using Kirchhoff's equations for the equivalent circuit shown in Fig. 5-53.

$$I_1 = I_2 + I_3 \qquad 2RI_2 = RI_3 \qquad I_2 = \frac{I_3}{2} \qquad I_1 = \frac{3I_3}{2}$$

$$2RI_1 + 2RI_2 = V_{REF} \qquad R_3I_3 + RI_3 = V_{REF}$$

hence

$$I_z - I_3 = \frac{V_{REF}}{4R} = \tfrac{1}{2} \, MSB$$

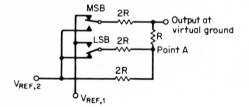

FIGURE 5-51   Operation of two-bit converter based on ladder network principle.

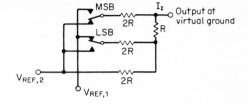

FIGURE 5-52   D/A converter with MSB operating.

Thus I_2 is now half the value observed in case 1. This approach can be repeated for any number of bits.

Since the ratio of resistor values need only be $2 \colon 1$, good temperature tracking can be achieved relatively easily. Furthermore, the ladder network has the advantage of constant output resistance. With reference to Fig. 5-51, the resistance at point A, looking back into the two $2R$ resistors in parallel, is R. From point B, this impedance is in series with R, giving a total of $2R$. Progressing toward the output mode, this $2R$ is parallel with another $2R$ gives R again. Repeating this for any number of bits will show that the impedance of the network at the output is a constant value, R, independent of the input code. This permits the addition of precision gain amplifiers to the converter to produce voltage outputs. The amplifier feedback resistor can be part of the resistor network to produce precise thermal tracking.

5-10-3   Recirculating A/D Converter

Figure 5-54 is but one example of a binary A-to-D converter based upon operational amplifiers and FET analog switches. The principle of operation is relatively simple:

1. The input V_A is sampled when switches S_1 and S_2 are closed and S_3 and S_4 are open.

2. The sample is applied to the amplifier A_2, which is in its reference mode.
   a. If V_A is greater than V_{REF}, the control logic output is a logic 1, and the amplifier will subtract V_{REF} from V_A and multiply the result by 2. This result is then recirculated via switches S_3 and S_2 and treated as a new input.

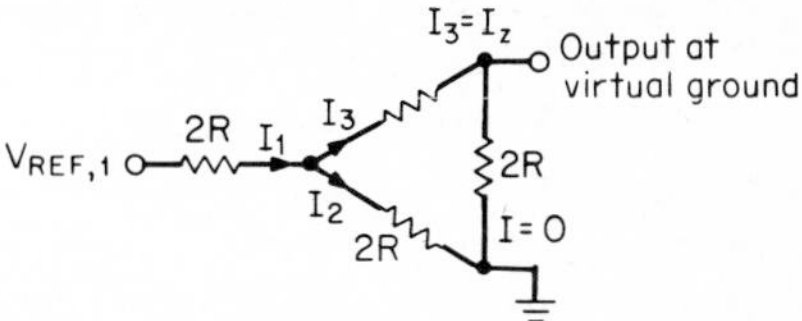

FIGURE 5-53   Equivalent circuit of Fig. 5-52.

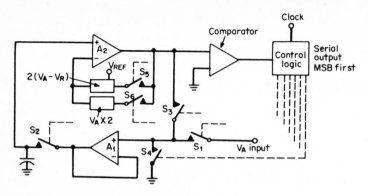

FIGURE 5-54 Recirculating A/D converter.

b. If the input is less than V_{REF}, the logic is a 0, and the amplifier will be switched to multiply the V_A signal by 2; this output signal is then recirculated as a new input. This would be repeated over a number of cycles up to the resolution capability of the unit, at which time S_1 will open to input a new sample.

The resolution of the converter may be improved by generating more cycles between signal sampling intervals. Switching spikes, which are created when the analog switches in the system are closed, can cause inaccuracies. The overall accuracy is critically dependent on the precision of the analog amplifiers.

The output will be in serial form but can, by the addition of a serial-to-parallel converter, be in parallel form. This design has been fabricated in monolithic PMOS integrated-circuit form. To compensate for MOS amplifier drift, the converter has an auto zero mode which is introduced prior to each input sample. In this mode, the input is grounded via S_4 (Fig. 5-54) and the output added to the analog amplifier reference for the next series of conversion cycles.

5-10-4 Deglitching

When the input digital code changes in a D/A converter, some switches turn ON and others turn OFF. The number of switches operated depends on the exact code change. However, since all types of switches have different turn-ON to turn-OFF responses, there will be rapid changes through many varying codes during the short transitional period. This results in transient spikes or "glitches" (Fig. 5-55). The magnitude of the glitch that can be tolerated depends on the overall system accuracy

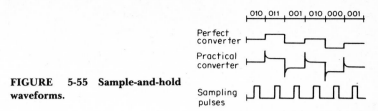

FIGURE   5-55   Sample-and-hold waveforms.

required. A low-pass filter would reduce the glitches, but for high accuracy, a sample-and-hold circuit is often placed after the converter output. The sample-and-hold switch is controlled by a pulse lying within the digit period, such that the sample is taken between transients, when the output is stable.

In circumstances where the input signal to an A/D converter may be changing by a value which is more than the least significant bit, during a complete conversion cycle it is possible to use a sample-and-hold before the converter to provide a stable input.

5-11   CONCLUSION

This chapter has presented some of the characteristics and applications of the FET as an analog switch. A large percentage of all FET applications are for switching. Several reasons for using a solid-state switch in place of a mechanical switch are obvious: high packing density, remote operation and no mechanical linkage, and reliability. These advantages apply to bipolars and FETs. FETs have several additional advantages:

○ High OFF to ON impedance ratios

○ Bilateral operation for large analog signal swing

○ Low power drive due to voltage-control action

Because of its widespread acceptance, the FET has been incorporated, as a switch, into many types of integrated circuits. For a more extensive treatment of this application of the FET, the reader is referred to a book available from Siliconix, Inc., *Analog Switches and Their Applications,* 1980.

6

VOLTAGE-CONTROLLED RESISTORS AND FET CURRENT SOURCES

6-1 The Nature of Voltage-Controlled Resistors
6-2 VCR Characteristics of FETs
6-3 How to Use the FET as a Voltage-Controlled Resistor
6-4 Signal Distortion: Causes
6-5 Reducing Signal Distortion
6-6 Linear Gain Control with VCR
6-7 Analysis of Feedback Linearized VCRs
6-8 FET as a Constant-Current Source
6-9 Cascade FET Current Source
6-10 Current-Regulator Diodes
6-11 Waveform Generation Using Current Regulators
6-12 Points to Remember when Choosing FETs as Current Sources

6-1 THE NATURE OF VOLTAGE-CONTROLLED RESISTORS

A voltage-controlled resistor (VCR) is a three-terminal device in which the resistance between two terminals is controlled by a voltage applied to the third terminal. The field-effect transistor is a device whose channel resistance is a function of its gate voltage; thus it can be used as a VCR. In this chapter we will describe the VCR characteristics and applications of n-channel JFETs. The principles can be applied to other types of FETs.

6-2 VCR CHARACTERISTICS OF FETS

Some of the unique resistance properties of the JFET can be deduced from the low-voltage output characteristic curves shown in Fig. 6-1a.

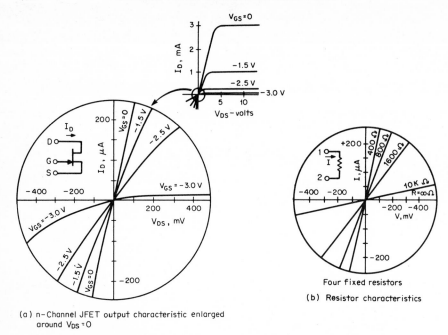

(a) n-Channel JFET output characteristic enlarged
around $V_{DS} = 0$

(b) Resistor characteristics

FIGURE 6-1   Comparison of FET and resistor characteristics.

The slope of the output curve V_{DS}/I_D is a function of V_{GS}; thus the drain-source resistance r_{DS} is controlled by V_{GS}.[1,3] For comparison the characteristics of four resistors are shown in Fig. 6-1 b. It can be seen that r_{DS} of the FET is also a function of V_{DS}; however, this effect is small *if V_{GS} or V_{DS} is small compared to $V_{GS\,(off)}$.*

The drain-source resistance with $V_{GS} = 0$ and with V_{DS} either at zero or at a very low value is given the symbol $r_{DS\,(on)}$. A graph of r_{DS} versus V_{GS} normalized to $r_{DS\,(on)}$ and $V_{GS\,(off)}$ is shown in Fig. 6-2a. Characteristics of typical devices with various geometry sizes and various values of $V_{GS\,(off)}$ are shown in Fig. 6-2 b. For a given geometry, units with high $V_{GS\,(off)}$ have the lower $r_{DS\,(on)}$. Units with the lower $r_{DS\,(on)}$ will also have the highest I_{DSS}. The relationship between these three parameters is approximated by the equation

$$r_{DS\,(on)} = -\frac{V_{GS\,(off)}}{2\,I_{DSS}} \qquad (6\text{-}1)$$

FETs specifically characterized for use as VCRs are offered by several manufacturers. Both n-channel and p-channel types are available. Figure 6-3 shows that $r_{DS\,(on)}$ versus $V_{GS\,(off)}$ relationship of two p-channel types

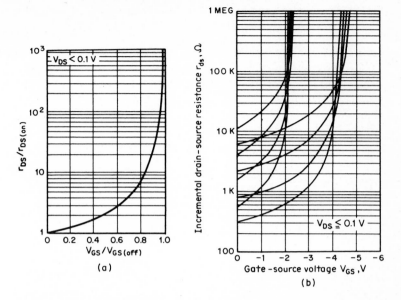

FIGURE 6-2 *(a)* **Normalized** r_{ds} **versus** V_{GS} **and** *(b)* **typical** *n*-**channel JFET characteristics.**

and three *n*-channel types offered by Siliconix Inc. These five device types cover a range of $r_{DS(on)}$ values from 20 to 8000 Ω. A FET does not have to be specifically characterized as a VCR to be useful in that function. For example, if a lower value of r_{DS} is needed, the type 2N5432, which has an $r_{DS(on)}$ range of 2 to 5 Ω, may be used.

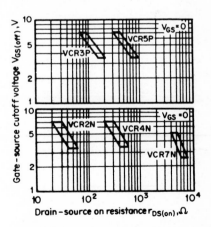

FIGURE 6-3 $r_{DS(on)}$ **(drain-source resistance at** $V_{DS} = V_{GS} = 0$**) varies as an inverse function of** $V_{GS(off)}$**.**

6-3 HOW TO USE THE FET AS A VOLTAGE-CONTROLLED RESISTOR

Figure 6-4 shows a simple voltage-divider attenuator circuit using a FET VCR to provide a means of voltage-controlling the attenuation. As $V_{control}$ is changed from zero to $V_{GS(off)}$ of the FET, r_{DS} changes from $r_{DS(on)}$ to infinity. Letting $g_{ds} = (r_{DS})^{-1}$, the output voltage is

$$V_{out} = V_{in} \frac{R_{load}(1 + g_{ds}R_{load})^{-1}}{R + R_{load}(1 + g_{ds}R_{load})^{-1}} \qquad (6\text{-}2)$$

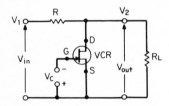

FIGURE 6-4 Simple attenuator circuit.

When V_{GS} approaches $V_{GS(off)}$, g_{ds} approaches zero and the FET causes no signal attenuation. If the load resistance is very large compared to R and $r_{DS(on)}$, Eq. (6-2) can be simplified to

$$V_{out} = V_{in} \frac{1}{1 + Rg_{ds}} \qquad (6\text{-}3)$$

In Chapter 2 [Eq. (2-8)] it was shown that

$$g_{ds} = g_{dso} \frac{V_{GS(off)} - V_{GS}}{V_{GS(off)}} \qquad (6\text{-}4)$$

Using this equation we can express V_{out} as a function of V_{GS}:

$$V_{out} = \frac{V_{in}}{1 + Rg_{dso}[V_{GS(off)} - V_{GS})/V_{GS(off)}]} \qquad (6\text{-}5)$$

Figure 6-5 shows the calculated V_{out}/V_{in} versus V_{GS} for the circuit shown in Fig. 6-4, assuming the constants indicated *and* assuming that the output is small compared to $V_{GS} - V_{GS(off)}$. For a given output voltage, as V_{GS} approaches $V_{GS(off)}$, nonlinearity increases and distortion increases.

6-4 SIGNAL DISTORTION: CAUSES

Figure 6-1a shows that r_{DS} increases as V_{DS} increases in a positive direction. This change in r_{DS} causes the distortion encountered in VCR cir-

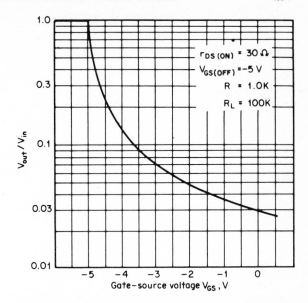

FIGURE 6-5 Calculated V_{out}/X_{in} for circuit of Fig. 6-4.

cuits; it occurs because the channel depletion layer is a function of V_{DG} as well as V_{GS} (see Figs. 1-7, 1-8, and 1-9). The output conductance g_{ds} approaches zero when *either* V_{GS} or V_{GD} approaches $V_{GS(off)}$. Also, when V_{DG} swings negative more than a few hundred millivolts, excessive gate current will flow and cause a dc offset voltage at the drain. This is due to the gate-drain *p-n* junction becoming forward-biased.

6-5 REDUCING SIGNAL DISTORTION

Distortion can be reduced and signal-handling capability increased by a simple negative-feedback technique. A portion of the drain signal is coupled to the gate as shown in Fig. 6-6. The desirable effect of this feedback can be explained as follows. If the gate-source voltage is fixed, a positive-going drain voltage tends to decrease the channel thickness,

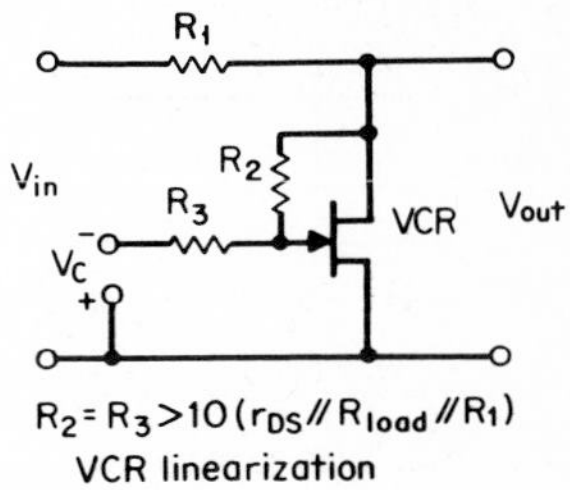

FIGURE 6-6 VCR linearization.

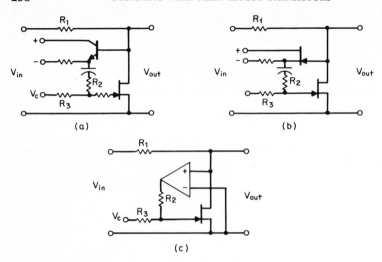

FIGURE 6-7   VCR feedback isolation with voltage followers.

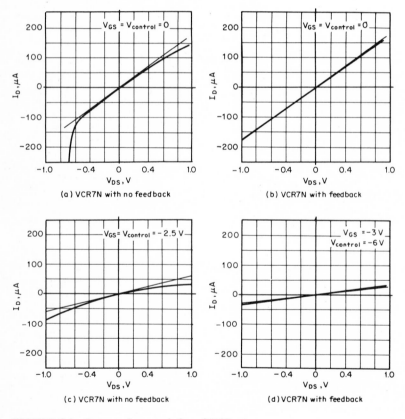

FIGURE 6-8   Output characteristics of VCR attenuators.

thus causing an increase in r_{DS}. By applying a part of the positive-going drain voltage to the gate, the gate-source voltage is made more positive (or less negative) and the thickness of the source end of the channel is increased. This compensates for the positive-going drain-gate voltage. A similiar effect occurs with a negative-going drain voltage.

This method of making V_{GS} a function of V_{DS} decreases the effect that V_{DS} has on g_{ds}, and thus distortion is reduced. The greatest improvement occurs when about $\frac{1}{2} V_{DS}$ is added to V_{GS}; thus in Fig. 6-6 $R_2 = R_3$. To avoid excessive loading of the signal, $R_2 + R_3$ should be much greater than R_1. If the dc control voltage must be blocked from the output, then a capacitor should be placed in series with R_2.

If the coupling of ac control signals from the gate to the output must be avoided, a voltage follower may be added to the feedback loop. Figure 6-7 shows three methods of doing this. Figure 6-8 compares the output characteristics of VCR circuits with and without feedback. It is obvious that linearity is greatly improved with feedback.

It will be noted that a sharp increase in I_D will occur if V_{GD} exceeds about 0.6 V. This drain-to-*gate* current results from the forward-biased drain-gate junction, not from a rapid change in *channel* conduction.

For comparison, a straight line representing a fixed resistor attenuator is superimposed on each of the output characteristics in Fig. 6-8.

6-6   LINEAR GAIN CONTROL WITH VCR

The simple attenuator circuit of Fig. 6-4 provides a V_{out} which has a reciprocal relationship with the control voltage V_{GS}. A linear control of gain can be achieved with the circuit shown in Fig. 6-9. This is a modified ordinary noninverting operational amplifier with feedback. The feedback is controlled by a FET VCR. The gain function of this circuit is

$$A_V = 1 + R_1 g_{ds} \tag{6-6}$$

where g_{ds} is the FET conductance ($= 1/r_{DS}$).

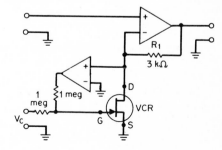

FIGURE 6-9   Amplifier with VCR gain control.

Using Eq. (6-4) we can express the circuit gain as a function of the control voltage V_{GS}:

$$A_V = 1 + R_1 g_{dso} \frac{V_{GS\,(off)} - V_{GS}}{V_{GS\,(off)}} \tag{6-7}$$

A minimum gain of unity occurs when V_{GS} is equal to or greater than $V_{GS\,(off)}$; the FET channel is pinched off and $g_{ds} = 0$. As V_{GS} is made less negative than $V_{GS\,(off)}$, g_{ds} and thus A_V will increase in a linear fashion. When V_{GS} reaches zero and goes positive, A_V will continue to increase. As has been pointed out before, when V_{GD} exceeds a few tenths of a volt positive, appreciable dc drain current will flow; therefore operating with positive V_{GD} or V_{GS} should be done with caution.

As was shown with the attenuator circuit, distortion in the gain-control circuit is reduced with drain-to-gate feedback. An op amp in the feedback isolates the output from the control V_C. The feedback circuit attenuates the control signal so that $V_{GS} = \frac{1}{2} V_C$. The voltage-gain equation for Fig. 6-7 can be written as

$$A_V = 1 + R_1 g_{dso} \left(1 - \frac{V_C}{2\,V_{GS\,(off)}} \right) \tag{6-8}$$

Figure 6-10 shows the calculated gain function for the circuit of Fig. 6-9. The $r_{DS\,(on)}$ and the $V_{GS\,(off)}$ of the FET VCR are assumed to be 30 Ω and -5 V.

If the application requires that the minimum gain be greater than unity, the FET may be shunted with another resistor as shown in Fig. 6-11. The factor R_1/R_2 is added to the gain equation, which becomes

$$A_V = 1 + \frac{R_1}{R_2} + R_1 g_{dso} \left(1 - \frac{V_C}{2\,V_{GS\,(off)}} \right) \tag{6-9}$$

Figure 6-12 shows several types of applications for VCRs. When choosing a FET to function as a VCR, there are several factors which should be considered:

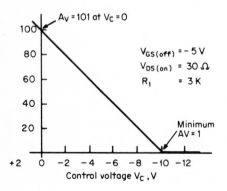

FIGURE 6-10 Amplifier gain vs. control voltage.

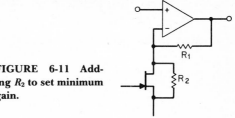

FIGURE 6-11 Adding R_2 to set minimum gain.

1. At low voltages the FET channel conducts equally well in either direction. No *p-n* junctions exist between the source and the drain; thus there is no inherent dc offset voltage.

2. The FET VCR behaves as a linear resistance only for small values of V_{DS}.

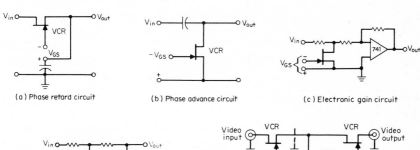

(a) Phase retard circuit

(b) Phase advance circuit

(c) Electronic gain circuit

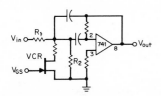

(d) Circuit increases overall dynamic range

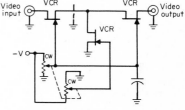

(e) Voltage controlled variable gain amplifier. Tee attenuator provides for optimum dynamic linear range attenuation

(f) Voltage-tuned filter octave range with lowest frequency at JFET $V_{GS(off)}$ and tuned by R_2 upper frequency is controlled by R_1

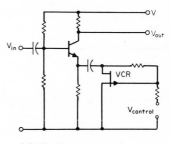

(g) Wide dynamic range gain control circuit

FIGURE 6-12 FET voltage-controlled resistor applications.

3. The channel conductance g_{ds} is approximately a linear function of V_{GS}.

4. The nonlinearity of r_{DS} increases as V_{GS} approaches $V_{GS\,(off)}$.

5. To improve linearity of the output characteristics and thus lower distortion, apply $\frac{1}{2}$ of V_{DS} to the gate with a feedback circuit.

6. FETs with high $V_{GS\,(off)}$ will have a larger dynamic range than those with low $V_{GS\,(off)}$.

6-7   ANALYSIS OF FEEDBACK LINEARIZED VCRs

The FET as a VCR is typically used in applications where V_{DS} is small and swings $\pm$ about zero. The V-I characteristics near the origin are shown in Fig. 6-1a. In this region, where V_{DG} is less than $-V_p$ and V_{GS} is less than V_p, I_D can be approximated by the quadratic function[2]

$$I_D = \frac{2 I_{DSS} V_{DS} (V_{GS} - V_p - \frac{1}{2} V_{DS})}{V_p^2} \tag{6-10}$$

Based on this approximation, the relation between distortion and control range maximum to minimum attenuation will be described. The simple attenuator circuit of Fig. 6-4 less R_L will be used. Most applications can be based on this simple example. The conductance at any point in the first and third quadrants of Fig. 6-1 is

$$G_{DS} = \frac{I_D}{V_{DS}} = -\frac{2 I_{DSS}}{V_p} \left(1 - \frac{V_{GS}}{V_p} \right) \tag{6-11}$$

$$-\frac{I_{DSS}}{(V_p)^2} V_{DS} = g_{ds} + \frac{g_{dso} V_{DS}}{2 V_P}$$

where g_{ds} is the differential conductance at the origin; when $V_{GS} = 0$, then $g_{ds} = g_{dso}$. The attenuation for the circuit of Figure 6-4 is

$$\frac{V_2}{V_1} = \frac{1}{1 + R G_{DS}} =$$

$$\left[1 + R g_{ds} + \frac{R g_{dso} V_1}{2 V_p \{ 1 + R g_{ds} + R g_{ds} V_1 / [2 V_p (1 + R g_{ds})] \}} \right]^{-1} \tag{6-12}$$

To reduce Eq. (6-12) to a more tractable form, the following inequality is introduced:

$$\frac{V_1 R g_{dso}}{2 V_p (1 + R g_{ds})^2} \ll 1$$

so that Eq. (6-12) can now be approximated by the expansion

$$V_2 = \frac{V_1}{1 + g_{ds}R}\left(1 - \frac{Rg_{ds}V_1}{2V_p(1 + Rg_{ds})^2} + \cdots\right) \qquad (6\text{-}13)$$

Only the second harmonic will be considered for the distortion, since the third is much smaller. For small distortion ($d \ll 1$ and $Rg_{dso} \gg 1$),

$$d = \frac{V_1 Rg_{dso}}{4|V_p|(1 + Rg_{ds})^2} \qquad (6\text{-}14)$$

If V_2 is held constant,

$$d = \frac{V_2 Rg_{ds}}{4|V_p|(1 + Rg_{ds})} \approx \frac{V_2}{4|V_p - V_{GS}|} \qquad (6\text{-}15)$$

Figure 6-13 shows a comparison of measured and calculated distortion. If V_{GS} approaches V_p, the above restrictions are violated; the expression for the distortion can no longer be applied. If V_{DG} becomes negative enough for appreciable I_G to flow (the gate-drain junction is forward-biased), the distortion will be higher than predicted. From Eq. (6-15) we get for a prescribed maximum distortion a maximum amplitude as a function of V_{GS}:

$$V_{2max} = 4d_{max}|V_p - V_{GS}| \qquad (6\text{-}16)$$

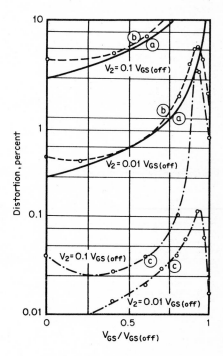

FIGURE 6-13   Distortion as a function of $V_{GS}/V_{GS(off)}$ for two different $V_2/V_{GS(off)}$: (*a*) theoretical; (*b*) measured; (*c*) measured with circuit of figure 26b (© 1968 IEEE; reprinted with permission, from Proceedings of the IEEE, Vol. 56, No. 10, October, 1968, pp. 1718–1719).

For a given d_{max} and V_{2max} the ratio of minimum to maximum attenuation is

$$\frac{A_{min}}{A_{max}} = m = \frac{1 + Rg_{DSS}}{1 + Rg_{DSS}\, V_{2max}/(4d_{max}|V_p|)} \approx \frac{4d_{max}|V_p|}{V_{2max}} \qquad (6\text{-}17)$$

valid only for $m > 1$. Note that the maximum distortion is reached only for minimum attenuation. Examples:

$$d_{max} = 10 \text{ percent } V_{2max} = 0.001\, V_p m = 400$$

$$d_{max} = \ \ 1 \text{ percent } V_{2max} = 0.01\, V_p m = 4$$

Although these relations are only first-order approximations, they give a good estimate of FET attenuator characteristics. The maximum amplitude is proportional to V_p. FETs with high V_p are desirable for attenuator applications. Unfortunately, the majority of commercially available FETs are made with low V_p for use in amplifiers.

There are several means of reducing distortion. By connecting two identical FETs in antiparallel or antiseries, nonlinearities can be cancelled out to a certain extent. A better linearization is possible using one FET with feedback. It has been shown above that the characteristics would be symmetrical if V_{GD} were the control voltage in the third quadrant. If $0.5V_{DS}$ is added to the control voltage, the two voltages V_{GS} and V_{GD} interchange when V_{DS} changes sign:

$$V_{GS} = V_H + 0.5V_{DS}$$
$$V_{GD} = V_H - 0.5V_{DS} \qquad (6\text{-}18)$$

Then Eq. (6-18) used in (6-12) gives

$$I_D = \frac{2I_{DSS}}{V_p^2} V_{DS}(V_H - V_p) \qquad (6\text{-}19)$$

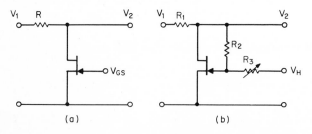

FIGURE 6-14 (a) **Controlled JFET attenuator and** (b) **controlled attenuator with "feedback" making characteristics linear and symmetrical.**

The resulting characteristic is linear and symmetrical about the origin. The improvement in distortion performance can be seen in Fig. 6-13. A distortion of 12 percent for $V_2 = 0.1V_p$ at $V_{GS} = 0.8V_p$ is reduced through linearization to 0.1 percent. Figure 6-14b shows a possible circuit. The frequency range of the controlled signal must be much higher than that of the controlling signal V_H to keep the direct interference of V_H on V_2 small. R_3 is set for minimum distortion. If V_2 and V_H are in the same frequency range, a high-impedance amplifier must be used. V_2 is at the input; the output is connected to the FET gate. The amplification is approximately 0.5 (adjustable). The control voltage is introduced through a second input so that no direct interference with V_2 occurs.

6-8   FET AS A CONSTANT-CURRENT SOURCE

The ideal "constant-current" source would supply a given current to a load independent of the voltage across the load. In such a case the output conductance of the current source would be zero. Also, ideally the current would be independent of temperature. The drain current of the JFET approaches saturation when the JFET is operated with the gate-drain voltage greater than $V_{GS(off)}$. Under this condition the output conductance g_{ds} is low, and the FET can be used as a current source.

Figure 6-15a illustrates a differential amplifier stage utilizing a FET as a constant-current source to provide a current I_S which is fairly independent of the common-mode voltage. The U401 dual FET is character-

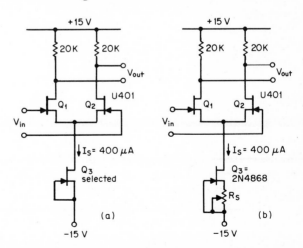

FIGURE 6-15   FET current source for differential amplifier.

ized at $I_D = 200$ μA; therefore the common-source current I_S should be set at 400 μA.

For example, we will assume that the Siliconix Inc. JFET geometry NS is used for the FET current source Q_3. If a device which has an $I_{DSS} = 400$ μA is selected from this geometry, then its gate can be shorted to its source as shown in Fig. 6-15a. The typical performance curves for the NS geometry indicate that the selected device would have $V_{GS(off)}$ less than 1 V and an output conductance of less than 1 μmho.

Figure 6-15b includes a resistor R_S in the source of Q_3. This permits a wide range of I_{DSS} values to be used and avoids the necessity of using a selected device. Q_3 could be a standard device such as the 2N4868, which is made with the Siliconix NS geometry.

The 2N4868 has an I_{DSS} range of 1 to 3 mA and a pinchoff range of -1 to -3 V. The approximate V_{GS} required for I_D is

$$V_G = V_{GS(off)}\left[1 - \left(\frac{I_D}{I_{DSS}}\right)^2\right] \qquad (6\text{-}20)$$

The source-resistor value can be determined by

$$R_S = \frac{V_{GS}}{I_D} \qquad (6\text{-}21)$$

In our example, then, R_S should be adjustable from 840 Ω to 7.4-kΩ.

In the circuit of Fig. 6-15a, if the common-mode input voltage is zero, Q_3 could be replaced with a 37.5-kΩ resistor (neglecting the V_{GS} of the input pair). With the -15-V supply, this would provide the desired 400-μA I_S. If, however, the common-mode voltage varies from, say, $+5$ to -5 V, then I_S would vary from 533 to 267 μA (again neglecting the small V_{GS} change of the input pair). In the case of the FET current source, assuming an output conductance of 1 μmho, the current I_S would change only from 405 to 395 μA.

In the above example with the source resistor, the ±5-V common-mode input signal would cause a 2.66-V peak-to-peak common-mode output. The use of the FET as a current source reduces the common-mode output to 0.1 V peak-to-peak. To achieve a similar low common-mode gain with a resistor, the source supply would have to be increased to -385 V so that the source resistor could be increased to 1 MΩ.

The output conductance g_{ds} of the FET is an important characteristic in its application as a current source. For a given geometry, g_{ds} is a function of I_D and an inverse function of V_{DS}. When adjusted to a given I_D, as in Fig. 6-15b, g_{ds} will be lower for FETs which have lower pinchoff voltage. For example, the performance curves for the Siliconix NS geometry indicate that units having 5 V, 3 V, and 2 V pinchoff will, respectively, have g_{ds} of 0.75 μmho, 0.58 μmho, and 0.42 μmho, when biased for an $I_D = 2$ mA.

These curves and typical curves for other geometries indicate that the g_{ds}-versus-I_D relationship has the form

$$g_{ds} = g_{dso} \left(\frac{I_D}{I_{DSS}} \right)^n \tag{6-22}$$

where g_{dso} and I_{DSS} are values at $V_{GS} = 0$. This form is similar to the g_{fs}-versus-I_D relationship. The value of the exponent n is a function of the device geometry and of the pinchoff voltage. For the various n-channel geometries evaluated, the exponent n lies between 0.33 and 0.9, with the higher-pinchoff units having higher values for n. However, the higher-V_p devices also have a higher g_{dso}/I_{DSS} ratio.

Long-channel geometries will typically have a much lower g_{ds}/I_D ratio than will short-channel geometries. For example, when biased to an $I_D = 1$ mA, a device made from Siliconix NS geometry has $g_{ds} = 0.35$ μmho. Another device with a similar V_p but made from the NRL geometry has $g_{ds} = 10$ μmhos at $I_D = 1$ mA. The NS is a long-channel geometry designed for low-noise, low-frequency amplifier application; the NRL is a short-channel geometry designed for VHF applications.

In the circuit of Fig 6-15b the output conductance of the current source will be lower than simply the g_{ds} of the FET. It is reduced by the degenerative feedback developed across the source resistor R_S. The approximate output conductance including the effect of R_S is

$$g_o = g_{ds} \left(\frac{1}{1 + R_S g_{fs}} \right) \tag{6-23}$$

Since the required value of R_S will be greater for units with higher V_P, the feedback will be greater. This gives some compensation for the higher g_{ds} of the high-V_P units.

In estimating the value of g_{ds} for a given device, it should be remembered that it is a function of both I_D and V_{DS}. The relationships of g_{ds} versus I_D and g_{ds} versus V_{DS} are functions of both the device geometry and $V_{GS(off)}$. It is probably easier to determine g_{ds} at a particular operating point by studying the performance curves given in the manufacturer's data book than to depend entirely upon a formula. Figure 6-16 shows typical performance curves for two of Siliconix Inc's n-channel JFET geometries. For both of these geometries it can be seen that g_{ds} is a function of I_D, V_{DS}, and $V_{GS(off)}$.

The rate of change of g_{ds} versus V_{DG} increases rapidly as V_{DG}

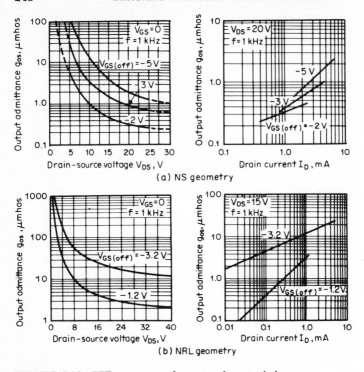

(a) NS geometry

(b) NRL geometry

FIGURE 6-16   FET output conductance characteristics.

approaches $-V_{GS(\text{off})}$. This indicates another advantage of units with low values of $V_{GS(\text{off})}$; they will perform well at low voltage.

The effect of temperature upon I_D was discussed in Chapter 2 (Sec. 2-10). For devices operating at I_{DSS}, such as Q_3 in Fig. 6-14a, the temperature coefficient (θ_I) of I_D will be a function of $V_{GS(\text{off})}$. If the pinchoff voltage is 0.65 V, θ_I will be near zero. Higher-pinchoff units will have a negative θ_I, and lower-pinchoff units will have a positive θ_I. The higher-pinchoff units can be biased to approximately $V_{GS} = V_{GS(\text{off})} + 0.65$ V to achieve zero θ_I. The use of a source resistor as in Fig. 6-14b can improve both the g_o and the θ_I of the current source.

6-9   CASCADE FET CURRENT SOURCE

Output conductance can be further reduced by cascading two FETs as shown in Fig. 6-17. In effect this circuit is similar to the current source of Fig. 6-15b, except that as a feedback element R_S is replaced with the output conductance of Q_1. The approximate output conductance

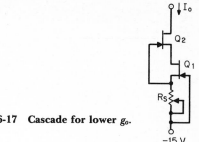

FIGURE 6-17   Cascade for lower g_o.

of the circuit of Fig. 6-17 is given by

$$g_o = \frac{g_{ds2}}{1 + (g_{fs2}/g_{o1})} \tag{6-24}$$

where g_{o1} is given by Eq. (6-23). In this circuit g_{ds1} will be higher than might be expected because V_{DS} of Q_1 is only equal to $-V_{GS}$ of Q_2.

6-10   CURRENT-REGULATOR DIODES

The desirable characteristics of the FET as a constant-current source stimulated the development of the current-regulator diode. One version includes an integrated source resistor to improve current control and to lower g_o and θ_J. Figure 6-18 shows the schematic, the symbol, and the equivalent circuit for this type of current-regulator diode. Examples of these units are Siliconix types CR022 through CR470. Three different geometries are used to cover the current range of 220 μA to 4.7 mA.

In Fig. 6-18 the diode I-versus-V characteristic curve is shown, and commonly used symbols and definitions are also given. Note that the drain end of the FET is called the anode and the source the cathode.

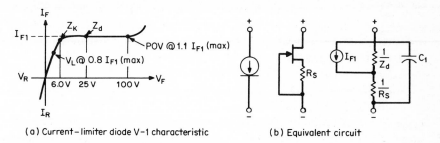

(a) Current – limiter diode V–1 characteristic   (b) Equivalent circuit

FIGURE 6-18   Current-regulator models and V-I characteristics (Siliconix CR022–CR470).

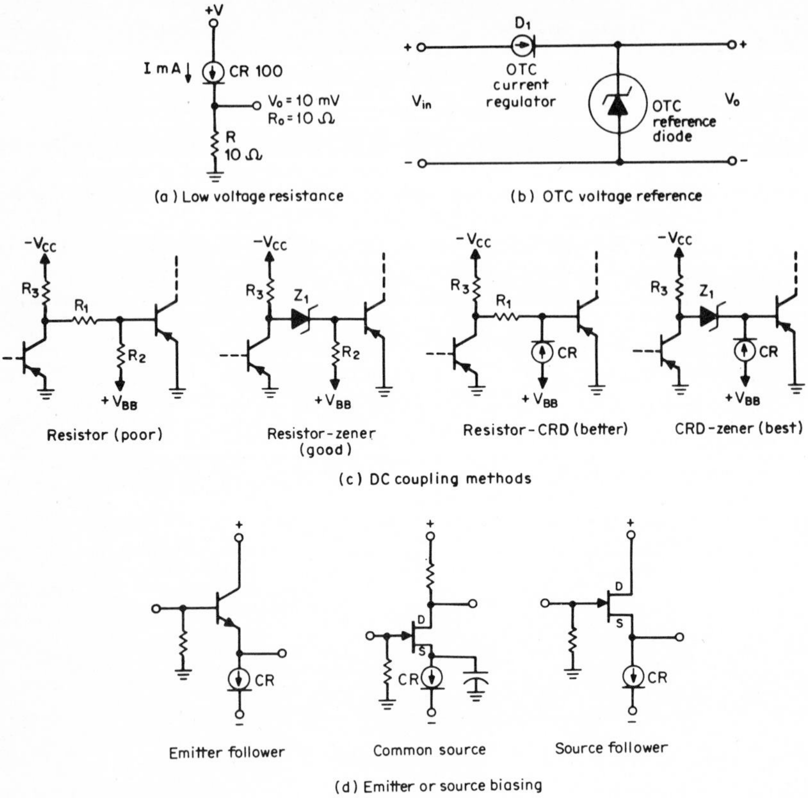

(a) Low voltage resistance

(b) OTC voltage reference

(c) DC coupling methods

Resistor (poor)   Resistor-zener (good)   Resistor-CRD (better)   CRD-zener (best)

Emitter follower   Common source   Source follower

(d) Emitter or source biasing

FIGURE 6-19 Current regulator applications.

The FET current-limiter diode is the electrical dual of the zener diode voltage regulator. The FET diode has low output conductance g_o, while the zener diode has low output resistance r_o.

Figure 6-19 shows several applications of the FET current-regulator diode. Some of the more important applications of current regulators include low-voltage references. Zeners are not available at voltages below 4 V. A precision voltage or millivolt reference is constructed in Fig. 6-19 where the current regulator simply drives a resistor. The output reference voltage is simply determined by $I_F R$.

Another benefit that cannot be obtained with zener references is the low noise of this voltage reference, constructed using the current regulator and a resistor. Neither of these devices is operating in a mode which contributes significant noise, especially at low frequency (which is hard to filter).

Consider the current regulator and resistor as a method of obtaining low-noise regulators even in the normal zener voltage range. One word of caution: the output impedance in this circuit (Fig. 6-19a) is determined by the resistor R.

An excellent solid-state zero temperature coefficient (0TC) voltage reference source is shown in Fig. 6-19b. Here we have combined an 0TC zener with an 0TC current regulator. Ripple from V_{in} to V_0 in Fig. 6-19b is reduced by more than 120 dB.

A very important application of FET current limiters is in dc coupling between gain stages and level shifting. Figure 6-19c shows methods of dc coupling between two transistor stages.

The use of the zeners and/or current regulators substantially reduces the gain loss otherwise encountered in resistive coupling dc amplifiers.

Differential amplifiers typically utilize current sources in the common emitters or common sources to help achieve the common-mode rejection of these amplifiers. Figure 6-15 shows how the FET current regulator is used in the differential amplifier. For the FET amplifier, common-mode rejection is

$$\text{CMRR}_{(V_{CM})} \approx 1 + 2\,g_{fs}\,Z_d \qquad (6\text{-}25)$$

where Z_d = output impedance of current source

g_{fs} = transconductance of the amplifier FETs

The output impedance of the current source is the key to improving common-mode rejection.

Single-ended transistor amplifier performance can be improved using current regulators. Figure 6-19d shows three different amplifiers using emitter and source current-regulator biasing. In the emitter follower the current regulator significantly increases the input impedance to the circuit and develops a gain closer to unity; less obvious is a lower transistor dissipation when supplying a heavy external load.

In the FET common-source amplifier the dc bias point is well defined using the FET current regulator. This allows zero temperature coefficient biasing of the amplifier. The same bias temperature stability is achieved in the source-follower circuit.

Another important application of the current regulator is in high-speed bootstrapping circuits. Figure 6-20 is a high-speed line driver using VMOS. In this circuit the current regulator CR047 minimizes power dissipation, at the same time maximizing speed. When Q_1 turns OFF, the CR047 discharges the parasitic capacitance rapidly in a controlled manner, turning ON Q_2 by enhancing the available gate voltage. Once the voltage across the current regulator drops below the limiting voltage V_L, it becomes very low resistance; thus the gate is tied to the top of the bootstrap capacitor.

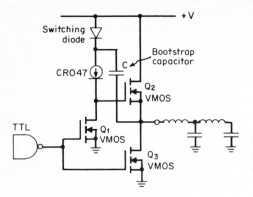

FIGURE 6-20   Application of current regulator (CR047) in a high-speed line driver.

6-11   WAVEFORM GENERATION USING CURRENT REGULATORS

A simple linear sawtooth generator can be constructed as shown in Fig. 6-21*a*. The use of the current regulator in series-opposing fashion makes possible the generation of high-quality triangular waves from a sine- or square-wave source, as Fig. 6-21*b* illustrates. Square-wave drive results in a better waveform at the zero crossings. Output frequency is identical to the input frequency.

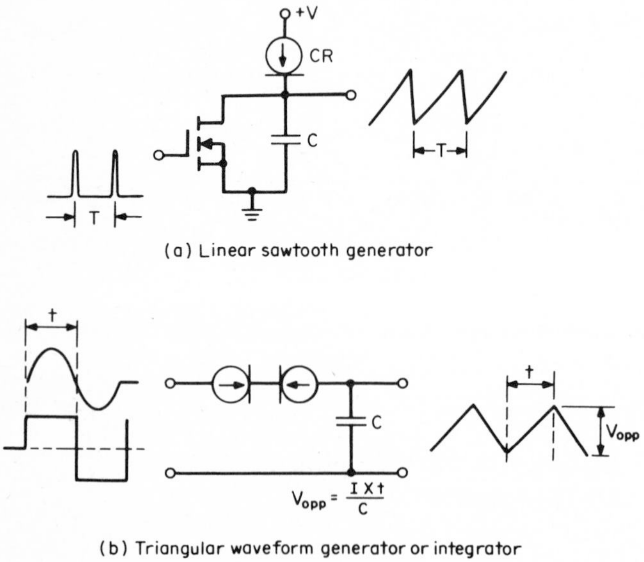

(a) Linear sawtooth generator

(b) Triangular waveform generator or integrator

$$V_{opp} = \frac{I \times t}{C}$$

FIGURE 6-21   Waveform generation.

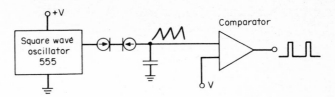

FIGURE 6-22   Pulse-width modulator.

A very useful application of the triangular-waveform generator is in pulse-width modulators, where the triangle wave is compared against a feedback voltage. The output of the comparator is an accurate pulse-width-modulated representation of the analog feedback voltage (Fig. 6-22).

A conventional clipper circuit or square generator is shown in Fig. 6-23a. One disadvantage is the poor dynamic resistance encountered in the low-voltage zener, which results in the fairly poor output waveform shown. The simple addition of series-opposing current regulators, shown in Fig. 6-23b, results in a flat-top output waveform and an efficient circuit. The output waveform is $\pm(V_z + 0.7)$ V.

As shown in Fig. 6-24, FET current-regulator diodes can be connected in series to extend the voltage range, in parallel to extend the current range, and in series-opposing to function as a bidirectional current limiter.

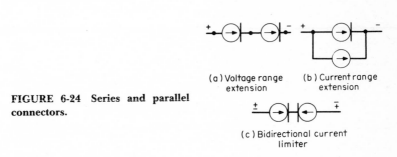

(a) Resistor zener   (b) Current regulator and zener

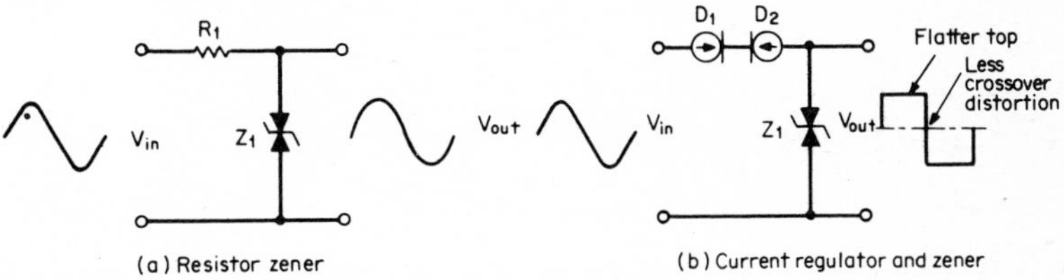

FIGURE 6-23   Square-wave generator or clipper.

(a) Voltage range extension   (b) Current range extension

FIGURE 6-24   Series and parallel connectors.

(c) Bidirectional current limiter

6-12 POINTS TO REMEMBER WHEN CHOOSING FETs AS CURRENT SOURCES

1. Shorting the gate to the source of a FET makes a simple current limiter.

2. Put a resistor in series with the source of a FET (Fig. 6-18) and you have a low-temperature-coefficient, high-output-impedance current source.

3. Place two FETs in a totem pole configuration (Fig. 6-17) and you increase the output impedance of the current source by more than an order of magnitude.

4. Applications requiring two-terminal current regulators without an internal supply are made very simply using depletion-mode FETs.

REFERENCES

1. Todd, Carl D., "FETs as Voltage-Variable Resistors," *Electronic Design,* pp. 66–69, September 13, 1965.

2. von Ow, H. P., "Reducing Distortion in Controlled Attenuators Using FETs," *Proceedings Letters,* IEEE, pp. 1718–1719, October 1968.

3. Watson, J., *Introduction to Field-Effect Transistors,* Siliconix, Inc., Santa Clara, Calif., p. 58, 1970.

BIBLIOGRAPHY

Gosling, W.: "Voltage Controlled Attenuators Using Field-Effect Transistors," *IEEE Trans Audio,* **AU-13,** pp. 112–120, September–October 1965.

Sevin, L. J.: *Field-Effect Transistors,* McGraw-Hill, New York, 1965.

Sherwin, J. S.: "Voltage Controlled Resistors (FET)," *Solid State Design,* pp. 12–14, August 1965.

Siliconix, Inc.: *FETs as Voltage-Controlled Resistors,* Application Note AN 73-1, Santa Clara, Calif., 1976.

7

POWER FETS

7-1   Introduction

7-2   The VMOS Technology

7-3   The Vertical JFET and DMOS Technologies

7-4   VMOS Characteristics

7-5   General Switching Applications

7-6   Drive Considerations

7-7   Temperature Considerations

7-8   Parallel and Series Operation

7-9   Amplifier Applications

7-10 RF Power

7-11 Summary

7-1 INTRODUCTION

Field-effect transistors, like their bipolar cousins, were for the first several years of their existence useful only at low (<1 W) power levels. While they possessed many theoretical advantages over their bipolar counterparts, the practical limitations in manufacturing high-power devices precluded FETs from competing with bipolar transistors and SCRs in power applications. The major limitation was that FETs were strictly horizontal devices; that is, their channels were parallel to the chip surface, so that their current densities were much lower than the bipolars' (which utilized vertical current flow). For a given current, the FET chip area had to be considerably larger, which meant a lower yield and higher cost. Medium-power FETs were therefore more costly to fabricate than their bipolar counterparts, while high-power FETs were nearly impossible.

Several new technologies have recently been developed to increase current density and allow production of high-voltage, high-current FETs. Three of these technologies—VMOS (vertical MOS), V-JFET (vertical JFET), and DMOS (double-diffused MOS)—are presently in production at various semiconductor companies. At the time of this writing (1980), VMOS devices with ratings of 12 A and 60 V and with ratings of 5 A and 400 V were available.

7-2 THE VMOS TECHNOLOGY

Figure 7-1 shows a cross section of a VMOS channel. The substrate, which becomes the drain and provides a low-resistance current path, is n^+ material. An n^- epitaxial layer (epi) increases the drain-source breakdown voltage by absorbing the depletion region from the drain-body junction, which is normally reverse-biased. Also, the epitaxial layer greatly reduces the feedback capacitance since the gate overlaps n^- rather than n^+ material.

A p^- body and n^+ source are then diffused into the epi, followed by the preferential etching of a V groove through the source and body and into the epi. Oxide is then grown and aluminum metalization deposited to form the source connection and gate. Finally, the chip is passivated (covered with glass) to keep contaminants from penetrating the gate oxide.

The processing, up to the point where the V groove is etched, is similar to that of the double-diffused epitaxial planar bipolar transistor, shown in Fig. 7-2 for comparison.

In operation, both the gate and drain are positive with respect to the source and body. The gate produces an electric field which induces an n-type channel on both surfaces of the p-type body facing the gate, allowing electrons to flow from the source, through the n-type channel and epi, and into the substrate (drain). Because current flow is entirely through n-type material, the VMOS is a majority carrier device similar to other types of FETs. A greater gate voltage enhances a deeper channel, so the current path from the drain to the source is wider and current flow is increased. For example, the VN66AF VMOS FET will conduct

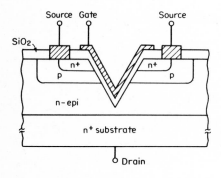

FIGURE 7-1 The cross section of a VMOS channel.

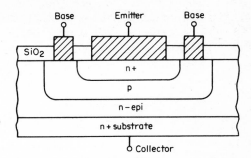

FIGURE 7-2   A double-diffused epitaxial planar transistor.

about 650 mA of drain current with 5 V between the gate and source,
and 2 A with 10 V gate-to-source.

Figure 7-3 shows a conventional horizontal MOSFET. The n^+ source
and drain are simultaneously diffused into the p-type substrate, which
also serves as the body. Current flows horizontally from source to drain
through the channel, which is induced on the top surface of the substrate.

The vertical structure of VMOS gives it several important advantages
over conventional MOSFETs:

1. The length of the channel is determined by diffusion
   depths, which are much more controllable than the mask
   spacings used to define the channel length of conventional
   MOS. With the shorter channel the width/length ratio of
   the channel—which determines current density—is greater.
   For example, the length of the VN66AF channel is about
   1.5 μm, while in a conventional MOSFET it is about
   5 μm.

2. Each V groove creates two channels, so current density
   is doubled for each gate stripe.

3. The substrate forms the drain contact, so drain metal is
   not needed on top of the chip. This further reduces chip
   area and keeps the saturation resistance low.

4. The high current density of VMOS results in low chip
   capacitance, especially the feedback capacitance (gate-

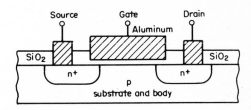

FIGURE 7-3   The cross section of
a conventional horizontal MOSFET.

drain), since the overlap of the gate and drain is kept to a minimum. Extra gate-drain overlap must be allowed in conventional MOSFETs to guard against mask misalignment.

7-3   THE VERTICAL JFET AND DMOS TECHNOLOGIES

Figure 7-4 is the channel cross section of the vertical JFET, another type of power FET which, like VMOS, has a high power density and high breakdown voltage capability. In the vertical JFET (V-JFET), current flows vertically from the drain through the channel between the gate fingers and into the source. A negative gate voltage $-V_{GS}$ causes the depletion region to reach further into the channel, reducing its width and constricting current flow. Ultimately, when the gate is sufficiently negative, the depletion regions from adjacent gate fingers touch, and current flow is stopped altogether. However, because the channel is short, it is not pinched off by increasing the drain voltage, as it would have been in a conventional longer-channel JFET.

The output characteristics of one type of JFET (2SK60) are shown in Fig. 7-5. The drain current does not saturate with increasing drain voltage but continues to be a function of drain voltage. The characteristics are similar to those of a triode vacuum tube; i.e., $\Delta I_D/\Delta V_{DS}$ is fairly large, and the voltage amplification factor g_{fs}/g_{ds} is quite low—on the order of 5.

The vertical JFET is necessarily a depletion-mode device (ON when $V_{GS} = 0$), while VMOS are typically enhancement-mode devices (OFF for $V_{GS} = 0$), although they could be made as depletion-mode devices.

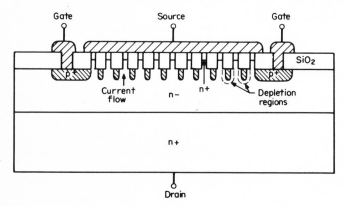

FIGURE 7-4   A cross section of a V-JFET.

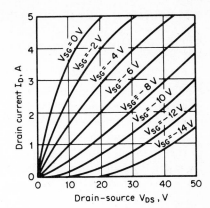

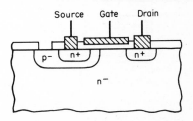

FIGURE 7-5 Output characteristics
of a 2SK60 vertical JFET.

A second difference is that the p^+ gate diffusion of the V-JFET has a higher resistance than does the aluminum gate metal used for VMOS, so the vertical JFET does not achieve the high-frequency response and fast switching times of which VMOS are capable. On the other hand, V-JFETs have a somewhat higher linearity than do other types of FETs, and are therefore suited for low-frequency amplifiers. To date, V-JFETs have primarily been used in high-quality audio power amplifiers.[5]

Double-diffused MOS (DMOS) is a first cousin of VMOS. It utilizes a short channel length and epi layer to achieve higher current densities with low capacitances. DMOS operation is fundamentally similar to that of VMOS, but it is a horizontal rather than a vertical device. A DMOS channel cross section is shown in Fig. 7-6. To date, DMOS devices have been used primarily at low and medium power levels, as high-speed switches or RF amplifiers. Because DMOS is a horizontal rather than a vertical structure, the current density is somewhat lower, and it is not expected to achieve the high power or high breakdown-voltage levels that VMOS will eventually attain.

Other power-FET structures are being developed;[6] so far all are variations of either VMOS, DMOS, or the vertical JFET. Time will tell which types are practical on a production basis.

FIGURE 7-6 A DMOS channel.

7-4 VMOS CHARACTERISTICS

The output characteristics of the VMOS type VN66AF, plotted in Fig. 7-7, are similar to those of a conventional MOSFET with these exceptions: The vertical scale is amperes rather than milliamperes, the output conductance g_{os} is low (the curves are flat rather than sloping) because of the buffering effect of the epi region, and the g_{fs} is constant (the lines are evenly spaced) above 400 mA. The constant g_{fs}, a characteristic of short-channel devices, is due to velocity saturation of the electrons in the channel. Above a certain threshold, increasing the electric field intensity does not increase the drift velocity. The g_{fs} of a conventional (long-channel) MOSFET, on the other hand, is proportional to the gate voltage; drain current is therefore proportional to V_{GS}.[6]

Figure 7-8 is a more graphic illustration of the transconductance vs. drain current for the VN66AF, showing the high linearity above 400 mA and the square-law characteristics below 400 mA.

Of the advantages that VMOS has compared to bipolars, many are well known in small-signal applications. Some that are apparent at high power levels are:

1. High input impedance and low drive current (typically less than 100 nA). The beta of a VMOS device (the output current divided by the input current) is therefore over 10^9. Since the resultant drive power is negligible, VMOS will directly interface to medium-high-impedance drivers such as CMOS logic or optoisolators.

2. No minority-carrier storage time. VMOS is a majority-carrier device—its charge carriers are controlled by electric

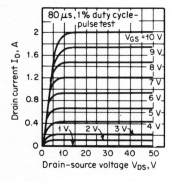

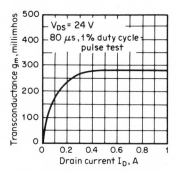

FIGURE 7-7 Output characteristics of the VN66AF.

FIGURE 7-8 Transconductance vs. drain current of the VN66AF.

fields, rather than the physical injection and extraction (or recombination) of minority carriers in the active region. The switching delay time is small, several nanoseconds, and is caused primarily by external parasitic elements (series gate inductance). The 2N6657, for example, is capable of switching 1 A ON or OFF in 4 ns, about 10 to 200 times faster than a bipolar.

3. No secondary breakdown or current hogging. Since the temperature coefficient of the VMOS drain current is negative (a bipolar's is positive), VMOS draws less current as the device heats up. If the current density tends to increase at one particular point of the channel, therefore, the temperature rises and the current decreases. The current automatically equalizes throughout the chip, so no hot spots or current crowding—which eventually leads to secondary breakdown in a bipolar—can develop. Similarly, current is automatically shared between paralleled devices so no ballasting resistors are needed.

7-5 GENERAL SWITCHING APPLICATIONS

The high input impedance and high speed of VMOS are desirable switch characteristics. VMOS will interface any driver capable of a 5- to 15-V swing to nearly any load requiring several amperes of current.

The basic switching performance of the VN66AF is shown in Fig. 7-9, while the corresponding test circuit is shown in Fig. 7-10; 40 pF is connected in series with the VN66AF gate to better match it to the 50-Ω source. The 2-ns turn-ON and turn-OFF delay is caused by the input capacitance charging and discharging through the equivalent series inductance of the package and test jig.

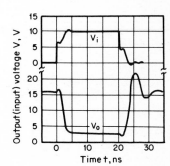

FIGURE 7-9 Switching performance of the VN66AF.

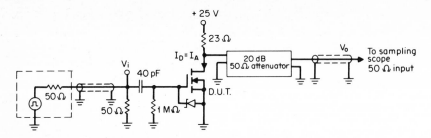

FIGURE 7-10 Switching test circuit for the VN66AF.

CMOS logic, such as the 34011 gate, may be used as a driver for the VN66AF in Fig. 7-11. A logic LOW to the input of the 34011 turns the VN66AF ON ($V_{GS} = 10$ V), while a logic HIGH turns the device OFF ($V_{GS} = 0$). The steady-state power dissipated by the circuit, exclusive of load current, is a maximum of 55 μW (0.15 μW typ).

Figures 7-12 and 7-13 depict the dynamic performance of the circuit shown in Fig. 7-11 when the load is 25 Ω. $V_{DD} = 15$ V when the logic supply voltage is 10 or 15 V and 7.5 V for $V_{CC} = 5$ V. The turn-ON and turn-OFF times when $V_{CC} = 10$ V are about 60 ns. Increasing V_{CC} to 15 V decreases t_{on} and t_{off} to 50 ns. Decreasing V_{CC} to 5 V increases the switching times to 120 ns. The input and Miller capacitances of the VN66AF present a load of 65 pF to the 34011 driver.

The switching time is decreased when several CMOS gates are paralleled to increase drive current to the VN66AF. For example, when four 4011 gates are paralleled and V_{CC} is 15 V, switching times are about 25 ns—most of it propagation delay through the 4011 (Fig. 7-14).

To further decrease switching times, additional peak drive current is needed to charge and discharge the gate input capacitance of the

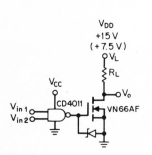

FIGURE 7-11 A CMOS gate driving the VN66AF.

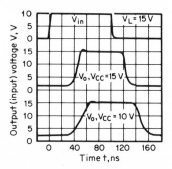

FIGURE 7-12 Switching performance of the 2N6657 driven by the CMOS gate.

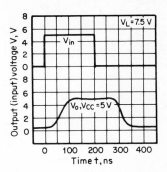

FIGURE 7-13 Switching performance of the VN66AF with a 5-V CMOS logic drive.

VN66AF. One solution is to use a MOS clock driver such as the MH0026. It is designed to deliver high peak currents into capacitive loads and to translate TTL levels into 15-V swings. Figure 7-15 is a typical switching circuit whose characteristics are shown in Fig. 7-16.

VMOS will also interface to standard TTL, but a pullup resistor is needed to ensure sufficient gate enhancement (see Fig. 7-17). If no pull-up resistor is used, the enhancement of the VMOS will be about 3 V, and the VMOS saturation current will be only about 200 mA. On the other hand, with 5 V provided to the gate by the pullup resistor, the VN66AF saturation current will exceed 500 mA—which is adequate for many applications.

If a higher saturation current or a lower ON resistance is needed, the gate drive voltage must be increased. Figure 7-18 shows how to use open-collector TTL with a 10- to 15-V pullup. Turn-ON time will be a function of the value of R_1, since it provides current to charge the input capacitance of the VN66AF. If a faster turn-ON time is needed, R_1 must be reduced, which may result in excessive power dissipation when the VN66AF is OFF. To solve this problem, use the totem pole

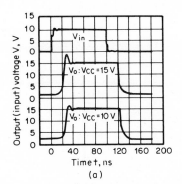

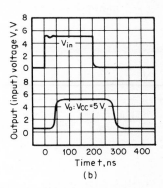

FIGURE 7-14 Switching performance of the VN66AF drive by four paralleled 4011 gates.

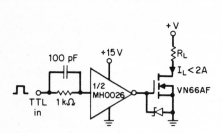

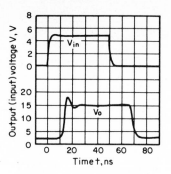

FIGURE 7-15   Driving the VN66AF with a MOS clock driver.

FIGURE 7-16   Switching performance of the VN66AF driven by a MH0026 MOS driver.

drive circuit shown in Fig. 7-19—it drives the VN66AF with an emitter follower with performance as shown in Fig. 7-20.

A second method of interfacing TTL to VMOS, a bipolar level shifter, is shown in Fig. 7-21. The 2N5130 amplifies the TTL output pulse and provides up to 15 V of enhancement to the VN66AF. The AM686 is a high-speed comparator, although other comparators—or for that matter, a TTL gate—could be used. For a faster turn-ON time than that shown in Fig. 7-22, or for a lower power dissipation in the OFF state, use the totem-pole driver.

Interfacing VMOS to ECL is not quite as straightforward, since ECL levels are inherently incompatible with VMOS drive requirements, but level shifting is still relatively easy. In Fig. 7-23, the VN66AF is used to increase the voltage and current capability of an ECL-compatible peripheral driver, the 75441. An alternative circuit, Fig. 7-24, uses discrete components to translate ECL levels into the 0- to 10-V swing

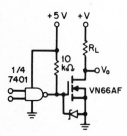

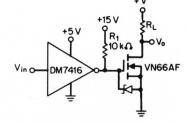

FIGURE 7-17   Driving the VN66AF with standard TTL.

FIGURE   7-18   Open-collector TTL is used to provide greater enhancement to the VN66AF.

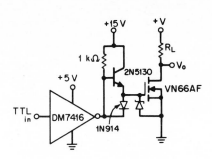

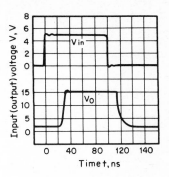

FIGURE 7-19 A "totem pole" drive increases switching speed and reduces dissipation.

FIGURE 7-20 Switching performance of the VN66AF driven by the "totem pole" open-collector TTL driver.

required for VMOS. Switching times of this circuit are less than 40 ns into 50 Ω.

When driving capacitive loads, such as cables or data buses, an active pull-up is required to deliver current into the load. The high-speed line driver shown in Fig. 7-25 uses a second VN66AF, with an inverter, to provide up to 3 A and switch 15 V across 1000 pF in less than 30 ns (Figure 7-26).

Ease of drive, ruggedness, lack of secondary breakdown, and fast switching speeds make VMOS well suited for switching power to a variety of loads, some of which are shown in Fig. 7-27. The fast switching of VMOS is especially helpful when designing switching regulators, since considerable power is lost while the switching element is traversing its active region. Figure 7-28 is the schematic of a 50-W, 200-kHz regulator using the VN64GA, a 12.5-A, 60-V VMOS device.[1] The regulator output

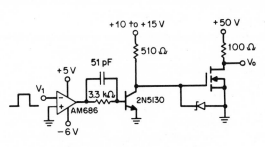

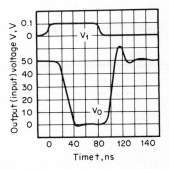

FIGURE 7-21 High-current interface.

FIGURE 7-22 Performance of the high-current interface.

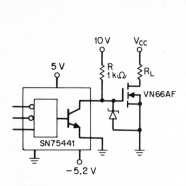

FIGURE 7-23 Using VMOS to buffer the output of an ECL-compatible peripheral driver.

FIGURE 7-24 A discrete ECL-to-VMOS interface circuit.

is 5 V at 10 A with ripple at less than 100 mV p-p. No output current limiting is included, although it may be added. Input supply is 28 V dc.

The 710 comparator acts as an oscillator, using L_1 as a reactive element and R_8 for hysteresis. C_5 couples the output ripple to the negative input of the comparator, where it is rejected as a common-mode signal. D_4, R_4, R_5, and C_3 form a bootstrap circuit which drives the gate 15 V more positive than the 28-V input. Six paralleled capacitors filter the output. The total impedance of one capacitor at 200 kHz is 0.05 Ω, and 0.01 Ω is needed to filter the 10-A peak-to-peak ripple current. Q_3 is the heart of a soft-startup circuit.

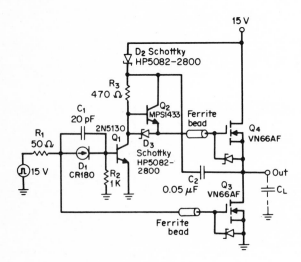

FIGURE 7-25 High-speed line driver.

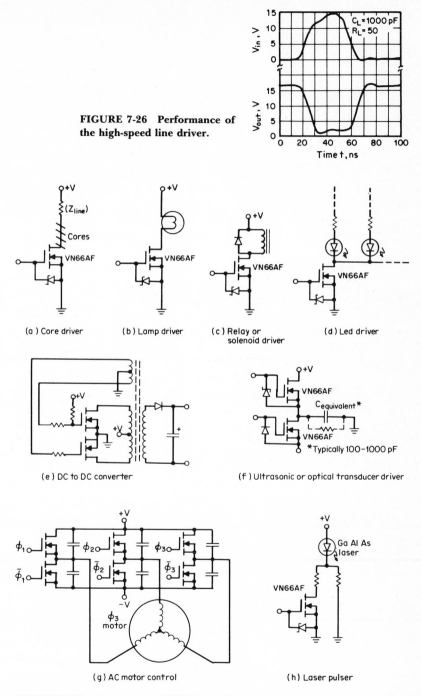

FIGURE 7-26 Performance of the high-speed line driver.

(a) Core driver

(b) Lamp driver

(c) Relay or solenoid driver

(d) Led driver

(e) DC to DC converter

(f) Ultrasonic or optical transducer driver

(g) AC motor control

(h) Laser pulser

FIGURE 7-27 Several typical VMOS applications.

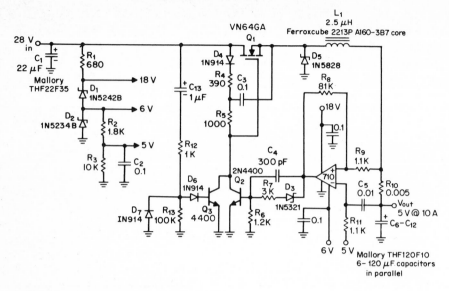

FIGURE 7-28 A 200-kHz switching regulator.

TABLE 7-1 A COMPARISON OF SWITCHING REGULATORS (28 V IN, 5 V AT 10 A OUT)

	20 kHz		200 kHz	
	Bipolar	*VMOS*	*Bipolar*	*VMOS*
Efficiency	82%	79%	72%	75%
Power output	50.0	50.0	50.0	50.0
Total input power	60.7	63.6	69.4	67.0
Fixed losses		4.85 W		
Drive power	0.17 W	0.44 W	1.4 W	0.87 W
Switching losses	1.9	0.55	9.6	3.7
Saturation losses	3.2	7.2	3.2	7.2
Ac core losses	0.06		0.2	
Dc coil losses	0.49		0.13	
Approximate recovery time for a 40% change in load	100 μs		10 μs	
Inductor core	3019 pot core ~0.85 in³, 1.2 oz		2213 pot core ~0.31 in³, 0.43 oz	
Capacitors	8 × 220 μF 1.0 in³		6 × 120 μF 0.45 in³	

Operation at 200 kHz, rather than the usual 20 to 25 kHz, has several advantages:

1. A smaller inductor, with lower dc (copper) losses, is used.

2. A smaller filter capacitor may be used.

3. The regulator responds faster to sudden changes in the load.

High-frequency operation does reduce the overall efficiency somewhat, because of switching losses, but not as much as it would in a comparable design using a bipolar or bipolar darlington transistor as the switching element. Table 7-1 compares bipolar and VMOS regulators operating at 20 kHz and 200 kHz. A circuit similar to Fig. 7-28 is assumed.

The high input impedance and linear transfer characteristic of VMOS make it easy to control either the average or the surge current to a load. Figure 7-29 is a simple light dimmer circuit which varies the average current into the light bulb by controlling the saturation current of the VMOS. R_1 and R_2 make the control of brightness more linear with the potentiometer shaft rotation. The disadvantage of this circuit is that the VN66AF operates in its linear region, so considerable power is wasted when the light is dimmed.

A more efficient method of varying the average current to the load is with pulse-width modulation (Fig. 7-30). The 4011 oscillates with a duty cycle which is determined by the ratio of R_1 and R_2, and drives the VMOS with 12-V pulses. Since the VN66AF is either fully ON or fully OFF, very little power is dissipated in the regulator itself.

A similar circuit can be used as an inexpensive audio alarm system. The 4011 gate provide a 2-kHz square wave to the VN66AF, which directly drives an 8-Ω speaker (Fig. 7-31).

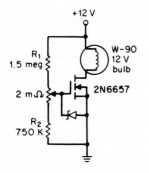

FIGURE 7-29 A linear light dimmer circuit.

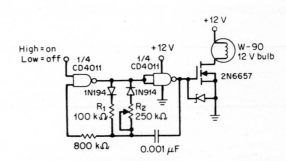

FIGURE 7-30 An efficient light dimmer circuit.

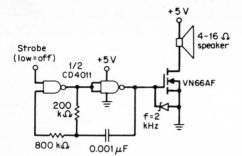

FIGURE 7-31   A 2-kHz audio alarm.

7-6   DRIVE CONSIDERATIONS

Many loads—motors and incandescent light bulbs, for instance—have an undesirably low impedance, resulting in high surge currents when power is first applied. The soft-startup circuits in Fig. 7-32 and 7-33 will minimize or eliminate these current surges, which in the case of an incandescent light bulb will increase life considerably. Adjust the 1-MΩ potentiometer in Fig. 7-32 until the desired maximum current is obtained, or use a fixed divider if a wider tolerance is allowable. R_1 and C_1 in Fig. 7-33 have a 0.1-s time constant to increase the drive voltage, and hence the drain current, of the VN66AF gradually.

While the 2N6657 can drive nearly any load of 2 A or less (3 A under pulsed conditions), the gate must be driven with a high enough enhancement voltage to support the required current. Refer back to Fig. 7-7: When the VMOS is driven by TTL providing a maximum V_{GS} of 5 V, the saturated drain current is 650 mA. The guaranteed worst case under these conditions is a drain current of 500 mA. If a minimum drain current of 1 A is required, a worst-case minimum of 10 V must be applied to the gate (6.25 V will typically be sufficient).

Applying more than the minimum enhancement voltage—15 V rather

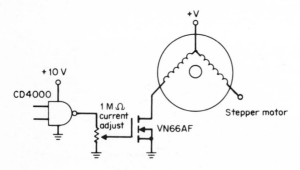

FIGURE 7-32   A circuit which current-limits the drive to the motor.

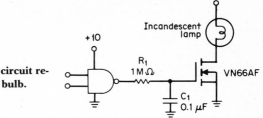

FIGURE 7-33 A soft startup circuit reduces the cold current of the bulb.

than 10, for instance—has two desirable effects: the ON resistance is reduced, and an extra margin of safety is provided to allow for the decrease in drain current as the VN66AF heats up. It is possible, as the drain current decreases with temperature, for the VN66AF to actually come out of saturation, which further increases dissipation.

7-7 TEMPERATURE CONSIDERATIONS

Typically, the saturated drain current of VMOS decreases 0.5 to 0.6 percent/°C due to the decrease in the mobility of electrons in silicon as temperature increases; likewise $r_{DS(on)}$ increases by the same amount (Fig. 7-34). If you assume a worst-case situation of 0.6 percent/°C, the ON resistance at a given temperature $[r_{DS}(T)]$ can be expressed in terms of the resistance at the ambient temperature $[r_{DS}(T_A)]$ by the expression

$$r_{DS}(T) = r_{DS}(T_A)e^{0.006\,\Delta T} \qquad (7\text{-}1)$$

where $\Delta T = T - T_A$, the rise in temperature.

This increase in ON resistance may lead to problems unless certain design precautions are taken. In a typical switching situation, such as that of Fig. 7-35, the current passing through the ON switch is nearly constant. In this example, 2 A of current passes through the VN66AF

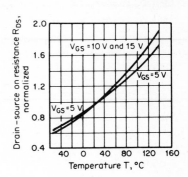

FIGURE 7-34 $R_{DS(on)}$ versus temperature of the VN66AF.

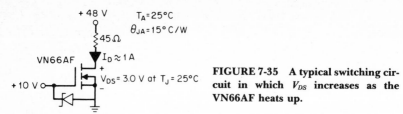

FIGURE 7-35   A typical switching circuit in which V_{DS} increases as the VN66AF heats up.

and causes it to heat up. As the ON resistance goes up, the voltage drop across the VN66AF increases, and the dissipation climbs further.

If inadequate heat sinking is used, the ON resistance and junction temperatures will increase until the resistance is stabilized by extra charge carriers which are thermally generated in the channel. Since this occurs above the maximum safe junction temperature of 150°C, and the long-term reliability may be impaired, it is desirable to anticipate this increase in ON resistance and temperature.

There are two ways to do this. The first, a rough rule of thumb, is to add an extra 50 percent to the actual power dissipation figure before calculating heat sink requirements. For example, if 1 A flows through a device whose ON resistance is 3 Ω at 25°C, the calculated power is 3 W. Now simply calculate the heat sink requirements using 4.5 W as the total dissipation, which will give a close approximation of the actual heat sinking required at moderate and high temperatures (it will be conservative if the temperature rise is slight). To calculate the heat sink requirements more precisely, express the rise in junction temperature ΔT in terms of the power dissipation and the junction to ambient thermal resistance θ_{JA}:

$$\Delta T = I^2 r_{DS}(T)\theta_{JA} \qquad (7\text{-}2)$$

which can be combined with Eq. (7-1) and rearranged as

$$\Delta T e^{-0.006\,\Delta T} = I^2 r_{DS}(T_A)\theta_{JA} \qquad (7\text{-}3)$$

This is the classical expression for the temperature risk of any fixed resistor, with the addition of the exponential term.

Solve Eq. (7-3) with the values shown in Fig. 7-35 to find the actual junction temperature.

$$\Delta T e^{-0.006\,\Delta T} = (1 \text{ A})^2 (3.0 \text{ }\Omega)(15°C/W)$$
$$= 45°C$$

Next, find ΔT in Fig. 7-36 by locating 45° on the vertical axis. The actual rise in junction temperature is located on the horizontal axis; it is 70°C, so T_J is 95°C.

Figure 7-36, a plot of $\Delta T e^{-0.006\,\Delta T}$ versus ΔT, is useful in finding the

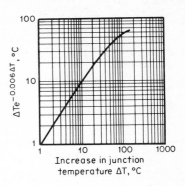

FIGURE 7-36 A plot of $\Delta T\, e^{-0.006\,\Delta T}$ versus ΔT is useful in finding the actual temperature rise of the VN66AF when the power dissipation at 25°C junction temperature is known.

actual temperature rise of the VN66AF when the power dissipation at 25°C junction temperature is known.

You can also use Fig. 7-36 to find the required heat sinking when the power dissipation and maximum allowable junction temperature are specified. For example, if the ambient temperature in Fig. 7-34 is increased to 50°C, $r_{DS}(T_A)$ becomes 3.5 Ω and the normalized power is 3.5 W. If the maximum junction temperature is specified as 125°C ($\Delta T = 75$°C), $\Delta Te^{-0.006\,\Delta T}$ must be less than 48°C and $\theta_{JA} < 13.7$°C/W. A heat sink with $\theta_{CA} < 5.4$°C/W should be used, since θ_{JC} of the VN66AF is 8.33°C/W. Note that if the rule of thumb were used and 50 percent added to the 3.5-W figure, the θ_{JA} would be calculated as 14.3°C/W—quite close considering the approximations involved.

7-8 PARALLEL AND SERIES OPERATION

If the required current exceeds the capability of one device, then several devices may be paralleled as in Fig. 7-37. No ballasting resistors or thermal matching networks are needed because the currents tend to equalize—if a particular device starts to draw more current, it heats up more and conducts less current than it would otherwise.

For example, an initial unbalance of ±20 percent (the typical worst-case figure) will reduce to ±14 percent if the junction temperatures

FIGURE 7-37 Paralleling the VN66AFs increases the maximum current-handling capability.

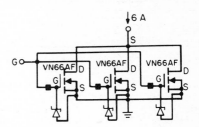

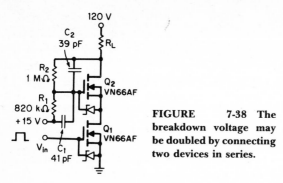

FIGURE   7-38 The breakdown voltage may be doubled by connecting two devices in series.

are allowed to approach their maximum limits. Because of the excellent high-frequency response of the VN66AF, ferrite beads or small-valued resistors ($\approx$100 to 1000 Ω) in series with each gate are necessary to suppress spurious high-frequency ($\approx$300 MHz) oscillations.

Devices may be connected in series to increase breakdown voltage, as shown in Fig. 7-38. R_1 and R_2 are larger because the drive current to the gate of Q_2 is small, while C_1 and C_2 form a capacitive divider which dynamically balances the gate drive and also ensures fast switching times by coupling charge to the gate of Q_2. C_1/C_2 should be approximately equal to R_2/R_1, with allowance for stray capacitance and the enhancement voltage of Q_2. The bottom of the divider chain is returned to +15 V, rather than ground, to ensure sufficient enhancement for Q_2 when the devices are ON. If resistor and capacitor values are properly selected, any number of VMOS may be series-connected in this manner.

7-9   AMPLIFIER APPLICATIONS

The constant-g_{fs} region of VMOS makes it well suited for linear applications. Distortion is low over a wide dynamic range when properly biased.

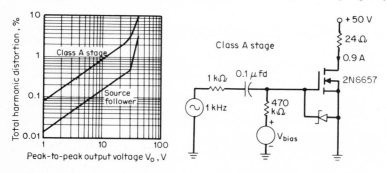

FIGURE 7-39   Harmonic distortion vs. voltage output for a simple class A stage and a source follower.

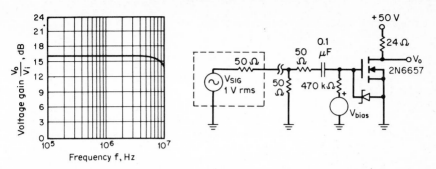

FIGURE 7-40 Frequency response of a simple class A stage.

Figure 7-39 is a plot of the harmonic distortion vs. output voltage for a simple class A test circuit employing the 2N6657, a 25-W VMOS. Distortion rises almost linearly with output voltage at low signal levels, but then rises more sharply as the positive signal peaks extend into the nonlinear g_{fs} region and the negative peaks saturate the device. The voltage gain of the circuit is about 6.5, equal to $g_{fs}R_L$ (0.27 mho $\times$ 24 Ω).

Using the 2N6657 as a source follower reduces the distortion by a factor of 5.5, which is slightly less than the amount by which the voltage gain is reduced. Figure 7-40 shows that the frequency response of a simple class A stage is flat to almost 10 MHz. The simple audio amplifier shown in Fig. 7-41 is equivalent to the audio output stage of many inexpensive radio and television receivers and phonographs. Power output is about 4 W from 100 Hz to 15 kHz. The design is greatly simplified by the use of an output transformer, and overall distortion is kept relatively low (2 percent at 3 W) by 10 dB of negative feedback. No thermal stabilization components are needed, since the drain current has a negative temperature coefficient.

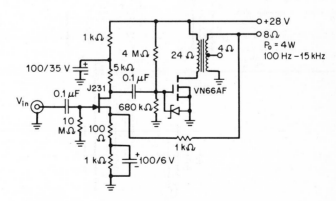

FIGURE 7-41 A simple audio power amplifier.

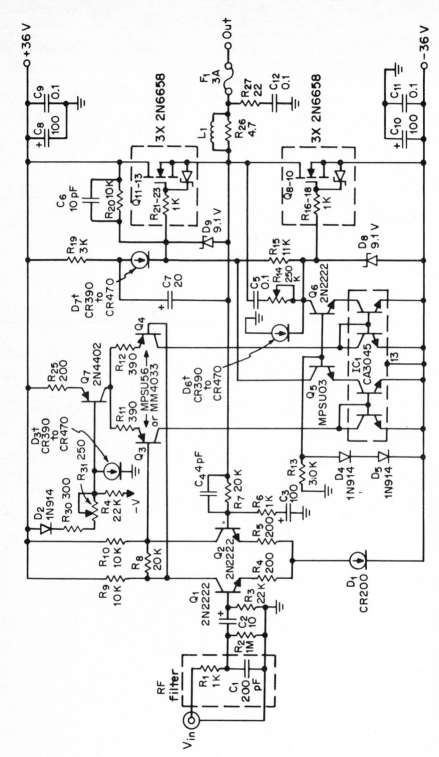

FIGURE 7-42 A high-quality, 40-W VMOS amplifier.

Figure 7-42 shows a high-fidelity 40-W audio amplifier suitable for high-quality stereo or quadraphonic systems.[2,4] This amplifier has low open-loop distortion. The relatively small amount of negative feedback (22 dB) and the good open-loop frequency response (400 kHz) minimize transient intermodulation distortion. Closed-loop frequency response (exclusive of the input filter) is flat to 4 MHz, and the slew rate is over 100 V/μs. The performance of the amplifier, which is operated class AB with an idling current of 300 mA, is shown in Fig. 7-43.

Since at the time of this design only n-channel VMOS devices were available, a quasi-complementary design was necessary, and some means was required to match the characteristics of the common-source and common-drain output stages. An effective method is to use a resistor (R_{15} in Fig. 7-42) to provide drain-to-gate feedback and a modulated current source in the common-source stage. Figure 7-44 details this technique and compares it to the corresponding circuit for an actual

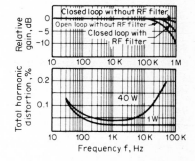

FIGURE 7-43 Gain and distortion vs. frequency of the VMOS amplifier.

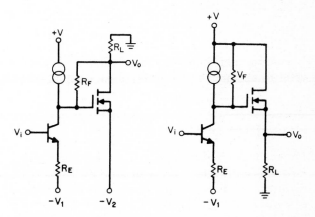

FIGURE 7-44 A quasi-source follower and a real source follower.

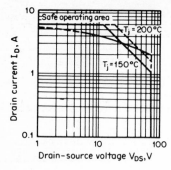

FIGURE 7-45 Current output vs. drain-to-source voltage of three 2N6658s in parallel, when $V_{GS} = 9$ V.

source follower. An analysis of the circuits reveals that both have the same values of gain and output impedance, which ensures a good match between the positive and negative waveforms during class AB operation. Exact matching is ensured by R_{14} and C_6 (Fig. 7-42). Functional output protection is provided by zener diodes, which limit the output current and device dissipation by limiting V_{GS}. Clamping V_{GS} at a maximum of 9 V limits the drain current to slightly less than 2 A at 25°C, less when the devices are hot. The resulting current limit vs. the drain-to-source voltage (Fig. 7-45) shows that short-circuit protection is possible when a 200°C maximum junction temperature is allowed for a brief interval (the time constant of the output fuse). The paralleled 2N6658s may be replaced with a single VN64GA, which was not available at the time of the initial circuit design.

7-10   RF POWER

VMOS has many advantages in RF power amplifiers and preamplifiers, including high gain, a high two-tone intermodulation intercept point, low noise, and the ability to withstand any VSWR. For further details, refer to Chapter 4, "High-Frequency Circuits."

7-11   SUMMARY

Power FETs typically rely on a short channel length and vertical current flow to increase current density and power capability. Their outstanding features, compared to bipolar transistors, include negligible dc drive current, extremely fast switching times, no minority-carrier storage time, a complete lack of secondary breakdown and current hogging, and low distortion. They are being designed into numerous power applications, including both general-purpose and high-speed switchers, high-quality audio amplifiers, and switching regulators.

REFERENCES

1. Shaeffer, L., "Improving Converter Performance and Operating Frequency with a New Power FET," presented at the Fourth National Solid-State Power Conversion Conference, May 1977. From "Proceedings of Powercon 4/Boston," pp. C2-1 to C2-8.

2. Shaeffer, L., "The MOSPOWER™ FET Audio Amplifier," Siliconix Design Aid DA76–1, May 1976.

3. Shaeffer, L., "Use FET's to Switch High Currents," *Electronic Design*, **9**:66–72, April 1976.

4. Shaeffer, L., "Vertical MOSFET's (VMOS) in High Quality Audio Power Amplifiers," presented at the 54th convention of the Audio Engineering Society, May 1976, AES Preprint No. 1106 (F-8).

5. Suwa, Hisashi, and Alciyasu Ishitani, "Vertical Field Effect Transistor and It's Application to High Fidelity Amplifiers," presented at the 51st convention of the Audio Engineering Society, May 1975, AES Preprint No. 1018 (F-7).

6. Yoshida, Kubo, and Ochi, "A High Power MOSFET with a Vertical Drain Electrode and a Meshed Gate Structure," *IEEE Journal of Solid-State Circuits*, **SC-11,** pp. 472–477, August 1976.

8

FETS IN INTEGRATED
CIRCUITS

8-1 Introduction
8-2 MOSFET Processes
8-3 Bipolar-FET Combinations

8-1 INTRODUCTION

Field-effect transistors are now a basic component in many integrated circuits, both analog and digital. While any type of FET—n- or p-channel, MOS or junction—can be fabricated on the same substrate with either devices of the same type or nearly any other kind of FET or bipolar device, some combinations are naturally more popular and useful than others. In this chapter we shall describe some of the more popular FET and FET-bipolar integrated-circuit basic processes, although many variations on each process have been developed to suit particular applications and manufacturing capabilities.

8-2 MOSFET PROCESSES

Figure 8-1 is an all-PMOS process, the earliest and simplest MOS process.[2,4] Only five masking steps are needed (p^+ diffusion, gate oxide, contact, metal, and oxide), so the cost is low and the yield high. Furthermore, the packing density (number of MOSFETs per unit area) is much higher than that of a bipolar process because no isolation diffusions are needed between devices. A high packing density reduces the cost of a function. The PMOS process is used primarily for medium- to high-complexity digital circuits, but has recently lost favor to NMOS.

NMOS, shown in Fig. 8-2, not only offers the same advantages of simplicity, low cost, and high packing density as PMOS, but it has a

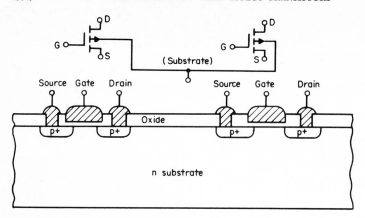

FIGURE 8-1　The PMOS process.

better speed-power product because negative charge carriers (electrons) have a higher mobility in silicon than do positive charge carriers (holes). This process was developed later than PMOS because it is much more sensitive to ionic contamination in the gate oxide, and only recently have processes been clean enough to ensure a high yield. However, NMOS has undergone extensive technical development and is a low-cost and widely used process for digital integrated circuits such as microprocessors and memory.

CMOS (complementary MOS), Fig. 8-3, is a combination of both n- and p-type MOSFETs. CMOS digital circuits dissipate very little power in the quiescent state because either the p or the n MOSFETs—but never both—are ON within a logic element, so there is no current flow except for a minimal leakage. At high switching frequencies (above sev-

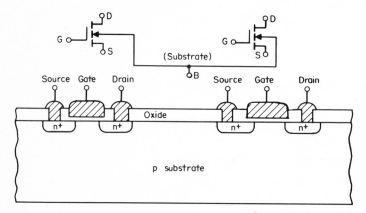

FIGURE 8-2　The NMOS process.

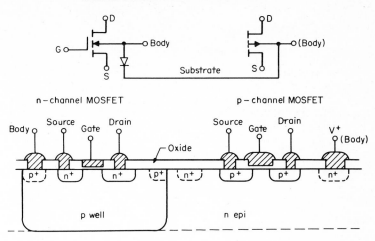

FIGURE 8-3   The CMOS process.

eral hundred kilohertz), however, the power dissipation becomes comparable to NMOS because circuit capacitances must be charged and discharged—which requires additional current. CMOS also requires more chip area for a given logic function because the n-channel MOSFET must be diffused into an isolated p-type well; this requires at least two more masking steps than PMOS or NMOS and is therefore more costly and difficult to fabricate. For digital circuits, CMOS is used when the complexity is low to medium; highly complex CMOS circuits are designed only when extremely low-power operation at low frequencies is needed—wristwatch circuits, for example.

CMOS is not confined purely to digital circuits, however—it is very popular for analog switches, operational amplifiers, and systems requiring both analog and digital circuitry. In an analog switch, the parallel combination of p- and n-channel MOSFETs exhibits a nearly constant

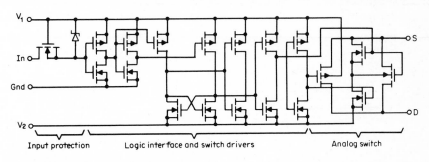

FIGURE 8-4   A CMOS analog switch.

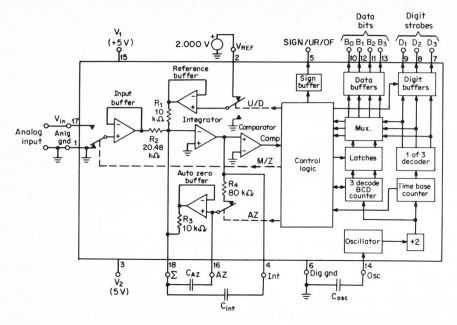

FIGURE 8-5 A monolithic CMOS chip containing both analog and digital circuitry.

resistance to any analog voltage between the positive and negative supply voltages. In a CMOS operational amplifier, the output is capable of excursions to within several millivolts of either supply rail, vs. several hundred millivolts or more for a more conventional bipolar design; a CMOS op amp also has several orders of magnitude lower input bias current than does a bipolar op amp, and both greater linearity and greater dynamic range than amplifiers made exclusively with either PMOS or NMOS. Figure 8-4 shows a CMOS analog switch system which

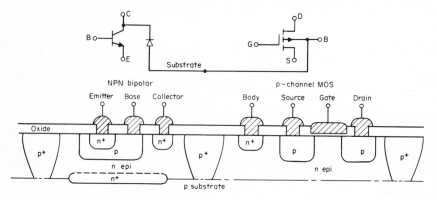

FIGURE 8-6 The bipolar-PMOS process.

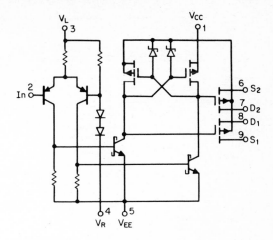

FIGURE 8-7   A bipolar-PMOS analog switch.

contains a digital interface and driver circuitry in addition to the analog transmission gate.[1]

Figure 8-5 is a block diagram of the LD130, a CMOS three-digit analog-to-digital converter. Notice that the buffers, integrator, and comparator are analog circuits (operational amplifiers), while the control logic and circuitry to the right of the logic is all digital. The analog switch functions utilize MOSFETs.[1]

8-3   BIPOLAR-FET COMBINATIONS

The bipolar-PMOS process (Fig. 8-6) is actually a forerunner of the previous MOSFET process; it was commercially developed in 1968 for

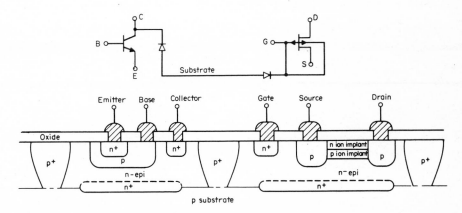

FIGURE 8-8   The BIFET (bipolar-JFET) process.

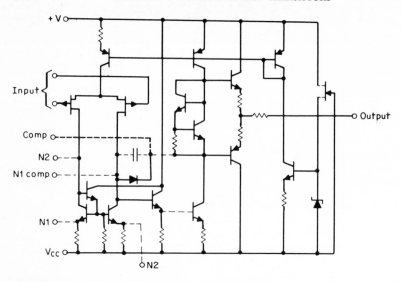

FIGURE 8-9 A BIFET op amp.

fabricating monolithic analog switch driver-gate combinations.[3] The only significant difference between this process and the standard planar bipolar process is an extra masking step for the PMOS gate, so the cost and complexity are only slightly greater. The bipolar-PMOS process is used in a variety of analog circuits—analog switches (Fig. 8-7), A-to-D converters which require MOSFET-input operational amplifiers and analog signal switching, and smoke-detector ICs which require a MOSFET-input comparator to interface with an ion-chamber smoke sensor.

Figure 8-8 is a versatile development—the BIFET (bipolar-JFET) process. Like the bipolar-PMOS process, it is basically a planar bipolar

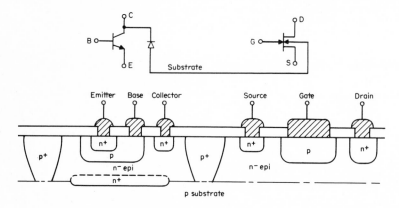

FIGURE 8-10 An *n*-channel JFET compatible with standard bipolar processing.

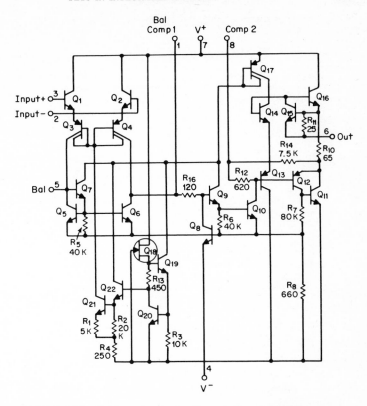

FIGURE 8-11 An integrated circuit operational amplifier using a JFET (Q_{18}) to ensure proper startup.

process, but with extra ion-implant processing steps to fabricate the channel of the p-channel JFET. This process is also used for a number of analog applications, including FET-input operational amplifiers (Fig. 8-9) and JFET analog switch/driver combinations.

The standard planar bipolar process can, with no extra processing steps, produce JFETs, but their parameters are difficult to control and close matching is nearly impossible. The process is shown in Fig. 8-10. JFETs of this type are used primarily for noncritical biasing and current sources in analog integrated circuits, especially since they are always ON when power is first applied and, therefore, ensure the startup of bipolar bias circuits (Fig. 8-11).

REFERENCES

1. "Analog Switches and Their Applications," Siliconix, Inc., Santa Clara, Calif., chaps. 1, 2, 1980.

2. Carr, W., and J. Mize, "MOS/LSI Design and Application," McGraw-Hill, New York, 1972.

3. Landsburg, G. F., "A Charge Balancing Monolithic A/D Converter," *Proceedings of the 1977 IEEE International Solid-State Circuits Conference,* pp. 98, 99.

4. Penney, W., and L. Lau, "MOS Integrated Circuits," Van Nostrand Reinhold Company, New York, 1972.

INDEX

Amplifiers:
  audio, 61
    VMOS, 274–278
  Cascode, 16, 48, 77–83
  common-drain, 63, 83–90
  common-gate, 63
  common-source, 8, 10, 26, 34, 55, 63, 145
  differential, 83, 89–92, 96–99, 102
    using current regulators, 251
  high-frequency, 16, 138
    broadband, 158, 160
    Cascode, 156–158
    common-drain, 155
    common-gate, 153–155
    common-source, 151–153
    design priorities, 140
    distributed, 140
    noise in, 22, 40, 152, 188–192
    problems of amplification, 139
    VMOS, 184–189, 278
  operational, 97, 101, 103–105
  R-C coupled, 10, 64
  voltage, 34, 63, 92
  VMOS, 186, 274–278
Analog switches (*see* Switches, analog)
Automatic gain control (AGC), 16, 48, 138, 153
Avalanche breakdown (*see* Breakdown)

Bias:
  constant current, 64, 75–77
    distortion, effects of, 116
    operating point, 64–65
  feedback, 84–86
  forward, 5–6, 8, 30, 54
  reverse, 5–6, 54
Bootstrapping, 79

Breakdown:
  avalanche, 7, 23, 29
  junction, 18, 46

Capacitance:
  abrupt junction, 11, 36
  feedback, 16, 139, 141
  feedthrough, 55
  gate, 58
  input, 58, 67, 139
  interelectrode: drain-body, 21
    drain-gate, 11, 21, 50, 139
    drain-source, 11, 37, 50, 139
    gate-channel, 18, 36, 47
  junction, 67
  output, 55
  parasitic, high-frequency effects of, 139
  switching transients, 219–226
Carriers:
  concentration of, 4
  density of, 12
  ionization of, 12
  minority, 21, 32
  mobility of, 4, 9, 12–13, 22, 26, 56
  velocity of, 9
Cascode circuits, 77–83, 145
  equivalent circuit, 81–82
  (*See also* Amplifiers)
Channel:
  conduction, 3–5, 13, 15, 18, 26, 56–57, 198
    (*See also* Conductance)
  diffusion, 20, 40
  length, 5, 9–10, 18, 20–21, 68
  resistance, 13
  thickness, 9, 15, 18, 20, 30, 41, 69, 95
  width, 18, 20, 68, 94–95

289

Characteristics of FETs:
  dynamic, 33
  output, 27, 30, 45–46, 68
  physical, 18–23
  in small-signal high-frequency applica-
      tions, 138
  static, 26–32
  transfer, 27, 45, 138
Charge transfer, 220–225
  deglitches, 230–231
  glitches, 224, 230–231
CMOS (complementary MOS), 222, 282–
    284
  switch, 205, 219
  (*See also* MOS)
CMRR (*see* common-mode rejection ratio)
Common-mode errors, 98–102
Common-mode rejection ratio, 100, 105,
    251
Common-mode voltage, 83, 92–93
Conductance, 13, 30
  channel, 4–9, 52
  drain-source, 31
  input, 141
  output, 8, 30, 141
Current:
  channel, 4
  constant source of, 64, 75–77, 83, 87, 91,
      245–248
  drain, 7, 10, 27
    saturation, 27, 41, 43–44, 81
  gate, 5–6, 29, 32, 54, 57, 104, 128–
      129
    breakpoint, 77–79, 104, 105
    leakage, 6, 12, 77, 111
  hogging, 21, 52, 261
  junction, 13
  regulators: applications of, 249–254
    FETs as, 249
Cutoff voltage:
  gate, 5, 27–28

Depletion, 7–8
  layer, 4, 7, 11, 57
  region, 8, 21
  thickness, 5, 12
  width, 4, 7, 37, 57
Derating factors, 58
Diffusion, 18
  channel, 20, 40
  gate, 40

Dissipation power, 62
  (*See also* Temperature)
Distortion:
  in analog switches, 210–211
  analysis of, 112–114
  in high-frequency mixers, 163–164,
      176–177
  rules for low, 121–122
  sources of, 114–120
  in VCRs (voltage-controlled resistors), 31,
      237–245
DMOS (double-diffused MOS), 255, 258–
    259
Drift, compensation for, 94–98
Duals (*see* Amplifiers, differential)

Epitaxial growth, 18, 20, 40, 69
Equivalent circuits:
  amplifier:
    Cascode, 82
    differential, 91, 99
      common-mode, 100
    high-frequency, 147
    voltage, 35, 64
  analog switch: CMOS control, 199
    dc, 193–197
    DG181, 201
    ON, 203
    JFET, 52
      control, 199
    sample-and-hold, 220–221
  charge transfer, 225
  converter, D-A, 229
  FET: admittance, 147
    capacitances, 37
    gate current, 79
    large signal, 162
    low-frequency, 36
    noise, 39
  multiplexers: leakage, 217
    two-channel, 224
  Tee isolation, 203
  VMOS (2N6659) ON/OFF, 208

Feedback, 16, 83–84, 86, 97, 139, 141
Feedthrough in analog switches, 199
  capacitance, 55
FET symbols, 17
Figure of merit, 55, 140
  (*See also* Gain, bandwidth)

Gain:
  bandwidth, 22, 157–158, 160
  conversion (mixer), 165–166
  midfrequency, 74
  power, 139, 141–142, 151
  voltage, 35, 81
Gallium arsenide, 22–23, 138
Geometries:
  distortion, effects of, 119
  mu, 141
  NC, 44, 68–70, 72
  NFA, 72
  NH, 72, 74
  NP, 72
  NRL, 72
  NS, 68–69, 72
  NT, 44
  NVA, 72
  NZA, 44, 72
  NZF, 72
  selector guide, 70
Glitch, 224, 230–231
Glossary of terms and abbreviations, 59

High-frequency admittance parameters,
    147–149
High-frequency amplifiers (*see* Amplifiers)
High-frequency circuits, 137–192
High-frequency mixers (*see* Mixers)
High-frequency oscillators, 181–185
High-frequency power FETs (*see*
    Amplifiers)
High-frequency power gain, 145–146
High-frequency VMOS, 184–188,
    278

Impedance:
  input, 6–7, 63, 143
  output, 143
Insulated-gate (*see* MOS)
Integrated circuits:
  bipolar-FET, 285–287
  CMOS, 282–284
  MOS, 281–285
Ion implant, 40, 69
Ionization:
  of carriers, 12
  impact, 78–79

JEDEC and house-numbered transistors:
  CRO39, 91, 93, 249
  CRO43, 107–108, 249
  J111–113, 38, 45
  M114, 54
  M116, 45
  U310, 37, 140, 149–150, 154, 159, 167,
    171, 183
  U311, 157
  U401, 91, 93, 107–108, 111
  U421, 111
  U431, 171
  2N3631, 45
  2N3822, 129
  2N3823, 137, 140, 168
  2N3970, 28–29, 32
  2N3971, 200
  2N4092, 38
  2N4339, 120
  2N4392, 38
  2N4393, 72
  2N4416, 72–74, 137, 140, 168
  2N4857, 38
  2N4867, 42, 119–120, 125
  2N4868, 28, 32, 42, 245–246
  2N4869, 42
  2N5397, 129, 140
  2N5432, 235
  2N5564, 38
  2N5638, 38
  2N6656, 51
  2N6657, 262, 269–270, 274–275
  2N6659, 207–210, 212
  3N140, 137
  3N201, 157
  VCR2N, 235
  VCR4N, 235
  VCR7N, 235
  VCR3P, 235
  VCR5P, 235
  VN66AF, 260, 262
  VN66AK, 212
  DG181, 201–203, 221–224
  DG181AP, 195, 202
  DG181BA, 200, 202
  DG172, 222
JFET (junction field-effect transistor),
    4–13, 52, 258
Junction:
  abrupt, 26, 36
  breakdown, 18, 46
  potential, 4

Metal-oxide semiconductor (*see* MOS)
Miller effect, 48, 79, 139, 156
Mixers:
 balanced, 168–181
 comparison of semiconductors, 169
 comparison of types, 169
 design criterion, 162–163
 double-balanced, 173–181
 high-frequency, 161–181
 interport isolation, 179
 single-balanced, 168–181
Mobility of carrier:
 temperature coefficient, 12–13
 (*See also* Carriers)
Models (*see* Equivalent circuits)
Modulation:
 cross-, 153, 163, 176–177
 inter-, 163, 176–177
MOS (metal-oxide semiconductor) or
  MOSFET, 3, 13–15, 18, 45–49, 54,
  255–279
 body, 15, 16, 18
 DMOS, 255, 258–259
 dual-gate, 16, 48–49, 82, 137–138, 145,
  157
 short-channel, 20, 49, 255–279
 switch, 204–205

Neutralization in high-frequency circuits,
  144, 152–153
Noise:
 amplifiers: bandwidth, 191
  matching, 190
 burst, 126–127
 characteristics, 38–40, 122–135
 current, 126, 128, 130–132, 161
 equivalent, 38–40
 figure, 123–124, 141
  mixer, 166
  temperature, correlation with, 188–
  190
 high-frequency, 22, 40, 152, 190–191
 impedance match, 152–153
 Johnson, 188–189
 local oscillator rejection, 178
 low JFET, 68, 105
 in operational amplifiers, 105
 oscillators, 182
 popcorn, 106, 126–127
 shot, 126
 test methods, 130–133

Noise (*Cont.*):
 voltage, 123–125, 130–132
 white, 131–132

Operational amplifiers (op amps) (*see*
  Amplifiers)
Oscillators, 181–185
Oxide, 3, 13, 18, 20, 256

Pairs, differential, 103–105
Parameters, high-frequency Z, Y, H, S,
  146–151
Phase compensation, 103
Pinchoff, 5, 7–8, 10, 18, 28
p-n, 3–7, 11, 14, 29, 55
Power FETs:
 DMOS, 255, 258–259
 types of, 255
 V-JFETs, 255, 258
 VMOS, 255–279
Power law, 27

Quality factor, 140–141

Resistance:
 gate-channel, 13
 input, 36, 77
 ON, 15, 52, 55
Resistors, voltage-controlled (VCRs):
 applications, 241
 characteristics of FETs, 233–235
  comparison to resistors, 234
 distortion in, 31, 236–245
 how to use JFETs, 236
 linearization: analysis, 242–245
  gain control, 239–242

Sample-and-hold circuit, 220–223
Saturation:
 drain current, 8–11, 27, 40
 velocity, 9, 260
Slope $(\Delta I_D/\Delta V_{DS})$, 7–8, 30, 33
Source follower, 83–90
Space charge, 32, 67
Stability:
 compensation, 103
 high-frequency: Linvill, 143
  Stern, 143

Switches, analog, 29, 32, 52, 193–231
  A-D and D-A, 226–231
  CMOS, 205–207
    leakage, 219
  comparison of types, 206
  current-mode, 193
  distortion in, 210–211
  feedthrough in, 199
  high-frequency, 199–200
  isolation, 200, 209, 213
  JFETs as, 197–199
  leakage characteristics, 215–219
  MOSFET, 204–205
  multiplexing, 193, 216
  selective summing, 194
  voltage-mode, 193
  VMOS, 207, 261–271

Temperature, effects of, 12–13, 56–58, 64, 94–96, 129, 271–273
Temperature coefficient, 12–13, 71, 129
Temperature ratings, 62
Threshold voltage, 15–16
Transconductance, 9–10, 27, 46, 142
  high-frequency mixers, 163–167
  small-signal, 27
Transfer characteristics, 12, 27, 45, 86
  effects on distortion, 114
Transients, switching, 219–226

Unilateralization, 144

VCRs (voltage-controlled resistors), 31, 233–245
Velocity:
  drift, 9, 23
  saturation, 9, 260
V-JFET (vertical JFET), 255, 258
VMOS (vertical MOS), 186, 207–214, 255–279
  applications, 264–269, 274–279
  breakdown voltage, 49–50
  capacitances, 50–51
  (*See also* Amplifiers)
Voltage:
  common-mode, 92–93
  effects on distortion, 117
  gain, 35
  gate: body to, 15
    channel to, 11, 13
    cutoff, 5, 27–28
    source to, 7, 11, 15
  noise, 123–125, 130–132
  offset, 55–56
  standing wave ratio (VSWR), 138
  threshhold, 15–16

Zener, 14